T0324211

GEOLOGY
OF
HIGH-LEVEL
NUCLEAR
WASTE
DISPOSAL

An introduction

GEOLOGY OF HIGH-LEVEL NUCLEAR WASTE DISPOSAL

An introduction

I. S. ROXBURGH
BSc, MPhil, PhD, DMS

LONDON   NEW YORK
Chapman and Hall

First published in 1987
by Chapman and Hall Ltd
11 New Fetter Lane, London EC4P 4EE

Published in the USA by Chapman and Hall
29 West 35th Street, New York, NY 10001

© 1987 Chapman and Hall

Printed in Great Britain by
J. W. Arrowsmith Ltd, Bristol

ISBN 0 412 29910 0

British Library Cataloguing in Publication Data

Roxburgh, I. S.
    Geology of high-level nuclear waste
    disposal—An introduction
    1. Radioactive waste disposal in the ground
    I. Title
    621.48'38   TD898.2

    ISBN 0 412 29910 0

Library of Congress Cataloging in Publication Data

Roxburgh, I. S., 1948–
    Geology of high-level nuclear waste disposal.
    Bibliography: p.
    Includes index.
    1. Radioactive waste disposal in the ground.
    2. Rock mechanics.   3. Hydrogeology.
    4. Engineering geology.   I. Title.
    TD898.2.R69 1987   628.4'4566   87–11713
    ISBN 0 412 29910 0

Contents

1

An introduction to high-level nuclear waste and the concept of geological disposal

1.1 CLASSIFICATION OF NUCLEAR WASTE

Waste can arise at a number of points in the nuclear fuel cycle starting with the mining and milling of uranium ore, during its refinement and enrichment, in the nuclear reactor itself and as a result of the reprocessing of spent fuel. Figure 1.1 shows the main stages in the nuclear fuel cycle while Table 1.1 shows the amount of waste arising at these various stages within the context of a 1 GWy(e) light-water reactor [1]. Table 1.2 shows the approximate theoretical quantities of some of the major fission products at different times after removal from a reactor that has operated at 1 MW(e), also for one year. The many types of waste involved may be classified according to their origin, with individual reactor types as shown in Table 1.3 giving rise to wastes with different detailed characteristics. Table 1.4 shows the varying nature of the radioactive liquid wastes arising from different reactor types. Wastes can also be classified by reference to their physical state and to their various levels of radioactivity and radiotoxicity. Knowledge of these various characteristics is required in order to assess the likely hazard posed by any particular waste and to determine the best method of long-term disposal.

The International Atomic Energy Agency (IAEA) has produced a fivefold classification of radioactive wastes on the basis of their radioactivity and heat output [4], and this is given in Table 1.5. This is a particularly useful method of labelling nuclear waste as it highlights its two main hazards, namely heat and radioactivity, and gives an immediate assessment of the degree of containment required and the time span involved. It is principally with the category I and II wastes that this book is concerned. Although the class III wastes also present a requirement for long-term isolation they do not have the problem of heat output associated with the class I and II wastes.

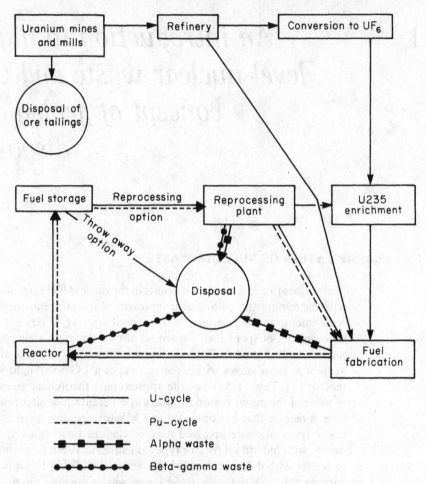

Figure 1.1 Major steps in the nuclear fuel cycle and main radioactive waste streams. (Redrawn from [1].)

1.2   ORIGIN OF CLASS I AND II WASTES

The wastes which comprise classes I and II wastes are principally those resulting from the irradiation of the nuclear fuel within a reactor and its subsequent reprocessing as reviewed by Flowers, Roberts and Tymons [5]. They comprise the fission products, the actinides, and the heavy elements formed in the fuel by various neutron capture and radioactive decay reactions. The largest volumes of highly radioactive and heat-generating wastes occur as a result of the reprocessing of spent irradiated fuel elements and are generally known as the high-level wastes (HLW). Uranium and plutonium are recovered from the spent fuel rods by solvent extraction using tributyl phosphate as

Table 1.1 Radioactive waste production in the nuclear fuel cycle. Solid or solidified waste equivalent to the generation of 1000 MWy(e) in a light-water reactor (modified from [1]).

Origin and type	Volume (after treatment and conditioning) (m^3)
1. *Uranium mining and milling*	
Ore tailings	60 000 (40 000 if Pu recycle)
2. *Fuel fabrication*	
UO_2 fuels	—
UO_2–PuO_2 fuels (for an annual reload of 500–700 kg of Pu)	10–50
3. *Light-water reactor*	100–500*
Various solid wastes and conditioned resins	
4. *Reprocessing*	3†
Solidified high-level waste (HLW)	
Compacted cladding hulls	3
Low- and medium-level β–γ solid waste	10–100
Solid and solidified α waste	1–10

*Depending on reactor type and conditioning process.
†From an original volume of liquid waste of about 15 m^3.

shown in Fig. 1.2. The high-level liquid waste (HLLW) remaining after reprocessing is a mixture of nitrate salts in nitric acid, containing about 99.9% of the non-gaseous fission products, about 0.1% of the uranium, less than 1% of the plutonium, and virtually all the other transuranium elements originally present in the spent fuel. It also contains activation products, corrosion products and chemicals such as cadmium, boron and fluorine added during reprocessing. The HLLW is allowed to boil for a while utilizing the heat generated by its own decay in an evaporative process to reduce its bulk. These HLLWs would then according to most disposal scenarios undergo a period of cooling prior to being solidified and confined in a geological repository.

Spent unreprocessed fuel (SURF) will also form class I and II wastes and this may also be referred to as HLW. If spent fuel were not reprocessed the amount of radioactivity to be disposed of would be greatly increased as reprocessing of spent fuel can with good technique reduce the levels of uranium and plutonium present to as little as 0.1–0.5% of the original uranium and plutonium present in the spent fuel.

Fuel fabrication and reprocessing can also give rise to other sources of radioactive waste, most notably from fuel element hulls and

Table 1.2 Approximate theoretical quantities of some major fission products, in TBq (kCi), at different times after removal from a reactor that has operated at 1 MW(e) for one year [6].

Radionuclide	At $t=0$		At $t=100$ days		At $t=5$ years	
^{85}Kr	7·07	(0·191)	6·92	(0·187)	4·88	(0·132)
^{89}Sr	1413·40	(38·2)	381·10	(10·3)		
^{90}Sr	52·91	(1·43)	52·54	(1·42)	44·40	(1·2)
^{90}Y†	52·91	(1·43)	52·54	(1·42)	44·40	(1·2)
^{91}Y	1809·30	(48·9)	536·50	(14·5)		
^{95}Zr	1820·40	(49·2)	629·00	(17·0)		
^{95}Nb†	1783·40	(48·2)	1061·90	(28·7)		
^{103}Ru	1143·30	(30·9)	219·04	(5·92)		
^{103m}Rh†	1143·30	(30·9)	219·04	(5·92)		
^{106}Ru	80·66	(2·18)	66·60	(1·8)	2·59	(0·07)
^{106}Rh†	80·66	(2·18)	66·60	(1·8)	2·59	(0·07)
^{131}I	932·40	(25·2)	0·15	(0·004)		
^{133}Xe	2046·10	(55·3)				
^{137}Cs	39·96	(1·08)	39·59	(1·07)	35·89	(0·97)
^{137m}Ba†	38·11	(1·03)	37·74	(1·02)	34·04	(0·92)
^{140}Ba	1912·90	(51·7)	8·51	(0·23)		
^{140}La†	1912·90	(51·7)	9·99	(0·27)		
^{141}Ce	1768·60	(47·8)	175·38	(4·74)		
^{143}Pr	1676·10	(45·3)	10·73	(0·29)		
^{144}Ce	987·90	(26·7)	769·60	(20·8)	9·99	(0·27)
^{144}Pr†	987·90	(26·7)	769·60	(20·08)	9·99	(0·27)
^{147}Nd	806·60	(21·8)	1·48	(0·04)		
^{147}Pm†	181·30	(4·9)	177·60	(4·8)	50·32	(1·36)

†Daughters.

components and storage-pond residues. These wastes are characterized by a high concentration of α (alpha) emitters and are generally known as α wastes though they may also be referred to as actinide or transuranium wastes. Where such wastes contain more than 10^{10} becquerel (Bq) per metric tonne of long lived α activity they are generally considered to require the degree of long-term disposal associated with geological disposal. The becquerel is now the adopted unit of radioactivity and is equal to one nuclear transformation per second, replacing the curie (Ci), with 1 Bq equalling 2.7×10^{-11} Ci. Table 1.6 gives the SI units used in defining and measuring radiation in its various contexts, and Table 1.7 lists the multiplying prefixes used to indicate multiples of the basic units.

Radiation is the transfer of energy through space, air or some other medium. To understand how this concept relates to radioactivity it is necessary to appreciate the nature of the atom and radioactive material. Everything is made up of atoms, which are composed of a nucleus containing almost the entire mass of the atom surrounded by negatively charged particles called electrons. The nucleus contains two principal particles, the positively charged proton and the

Table 1.3 The features of some nuclear reactors [2].

Reactor type	Fuel	Moderator	Control rods	Coolant
Thermal				
Magnox	Uranium metal in Mg alloy cladding	Graphite	Boron	CO_2 under pressure
CANDU	Uranium metal in stainless steel cladding	Heavy water (D_2O)	Boron in steel	Heavy water under pressure
Gas-cooled (GCR)	UO_2 enriched to 2·5% ^{235}U in zirconium alloy cladding	Graphite	B_4C	CO_2 under pressure (or He)
Pressurized water (PWR); also (LWR)	Uranium metal enriched 2–5% ^{235}U in stainless steel cladding	Water* at high pressure	Ag/In/Cd or B_4C	Water
Boiling water† (BWR)	UO_2 enriched to $\geqslant$1·5% ^{235}U	Water under pressure	B_4C	Water
Breeder				
Dounreay (DFR)	Uranium metal enriched to 45·5% ^{235}U in niobium cladding	None	B_4C	Molten Na/K alloy
Prototype (PFR)	Uranium and plutonium oxides	None	B_4C	Molten Na

*Water refers to H_2O unless specified as heavy water D_2O.
†BWR uses steam from coolant directly to drive the turbines.

Table 1.4 Nature of radioactive liquid wastes from different reactor types [3].

Type of reactor	Fuel				Characteristics of waste solution entering evaporator				Initial characteristics of concentrate to be stored			
	Type of fuel	Type of cladding	Typical burn-up (MW d t⁻¹)	Minimum cooling time before reprocessing T	Volume (1 l t⁻¹ heavy metal)	Activity at time T (Ci l⁻¹)	Heat content at time T (W l⁻¹)	Possible concentration factor	Volume after evaporation (l t⁻¹)	Activity at time T (Ci l⁻¹)	Heat content at time T (W l⁻¹)	Acidity N
LWR (USA)	UO₂	Zircaloy 2 or 4	29 000*	3 years*	5 200	150	0·6	14	380	2100	8·3	4–7
LWR (France)	UO₂	Zr	33 000	1 year	9 800	230	0·93	20	540	4100	18·5	2·5
LWR (UK)	UO₂	Zircaloy	33 000	150 days	6 250	860	3·2	16·5	400	14000	5·3	0·5–1·0
LWR (India)	UO₂	Zircaloy	15 000	150 days	7 200	300	0·9	9	800	3000	9·0	2–3
LWR (Japan)	UO₂	Zircaloy	28 000	180 days	5 500	210	0·59	16	350	3100	9·3	2·5
VVER (USSR)	UO₂	Zr	28 000	3 years	5 500	60	0·27	13	420	730	3.3	4–6

Reactor	Fuel	Cladding									
Gas-cooled (France)	U–Mo	Mg	3 000	1 year	7 600	70	0·24	70	110 4800	17	0·8
Gas-cooled (France)	U/Si/Al	Mg	5 000	1 year	5 400	100	0·34	50	100 5400	18	2·5
Magnox	U (Nat.) Magnox		1 300 3 500	125 days	4 500	460	0·25	50 100	90 45	12	3
AGR	UO₂	SS	37 000 18 000	1 year	5 000	600	0·33	20	-100	16	
CANDU			No plans for reprocessing								
HTGR	UC₂/ThC₂	Graphite, SiC	100 000	180 days	5 700 1700		7·5		3 600	3 600	
PFR	(U, Pu)O₂	SS	60 000	180 days	9 100 1200‡ 900†		2·5		4† 5‡	3 600	10§ 12§ 3
TR	U–Al	Al	200 000 400 000	~1 year	400 000 200		1·3		0·8	300 000	1 -1

*Probable equilibrium values for USA (currently there is no reprocessing of commercial nuclear reactor fuel).
†1st cycle, mean of outer and inner.
‡6th cycle, mean of outer and inner.
§Assuming evaporation immediately (in practice, evaporation will be delayed).

Table 1.5 General characteristics of waste categories with regard to disposal [4].

Waste category	Important features*
I. High-level, long-lived	High β/γ Significant α High radiotoxicity High heat output
II. Intermediate-level, long-lived	Intermediate β/γ Significant α Intermediate radiotoxicity Low heat output
III. Low-level, long lived	Low β/γ Significant α Low/intermediate radiotoxicity Insignificant heat output
IV. Intermediate-level, short lived	Intermediate β/γ Insignificant α Intermediate radiotoxicity Low heat output
V. Low-level, short-lived	Low β/γ Insignificant α Low radiotoxicity Insignificant heat output

*The characteristics are qualitative and can vary in some cases; 'insignificant' indicates that the characteristic can generally be ignored for disposal purposes.

neutron. While weighing almost the same as the proton the neutron has no charge. The number of protons in the nucleus is fixed and gives the atomic number of any particular element. The number of neutrons in the nucleus, however, may vary. As the atomic number increases the number of neutrons increasingly exceeds the number of protons, and the nuclei become unstable and disintegrate to achieve stability. This process of disintegration gives rise to radiation in the form of α and β (beta) particles and γ (gamma) rays. The process by which atoms emit α and β particles and γ rays is called radioactive decay and the energy emitted is collectively known as ionizing radiation.

While it is not possible to say at what time any particular atom of an element will decay, it is known that radioactive decay occurs at a rate proportional to the number of atoms. This allows for the calculation of what is known as the half-life of an element, that is the time it will take for half the material to have decayed. For example, a gram of radium-226 with a half-life of 1600 years will have reduced to half a gram in 1600 years, a quarter of a gram in 3200 years and an eighth of a gram in 4800 years.

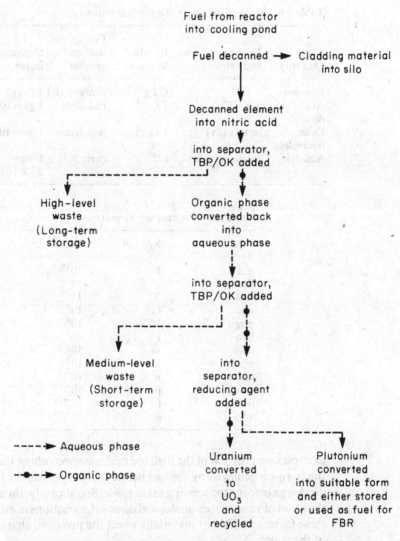

Figure 1.2 Flow diagram of the reprocessing of irradiated nuclear fuel. TBP = tributylphosphate, OK = odourless kerosene [6].

1.3 AMOUNTS OF WASTE INVOLVED

The total amount of class I and II wastes arising will be determined by the numbers and types of reactors in operation at any one time, the degree of spent fuel reprocessing, the possible introduction of breeder reactors and the recycling of plutonium. The precise composition of HLW as a function of time can be calculated from a knowledge of the

Table 1.6 SI units for use in the nuclear industry.

Quantity	New named unit and symbol	In other SI units	Old special unit and symbol	Conversion factor
Exposure	—	$C\,kg^{-1}$	röntgen (R)	$1\,C\,kg^{-1} \sim 3876\,R$
Absorbed dose	gray (Gy)	$J\,kg^{-1}$	rad (rad)	$1\,gy = 100\,rad$
Dose equivalent	sievert (Sv)	$J\,kg^{-1}$	rem (rem)	$1\,Sv = 100\,rem$
Activity	becquerel (Bq)	s^{-1}	curie (Ci)	$1\,Bq \sim 2 \cdot 7 \times 10^{-11}\,Ci$

Table 1.7 Prefixes for SI units.

Prefix	Symbol	Factor
exa	E	10^{18}
peta	P	10^{15}
tera	T	10^{12}
giga	G	10^{9}
mega	M	10^{6}
kilo	k	10^{3}
milli	m	10^{-3}
micro	μ	10^{-6}
nano	n	10^{-9}
pico	p	10^{-12}
femto	f	10^{-15}
atto	a	10^{-18}

isotopic composition of the fuel, the neutron spectrum of the reactor, the burn-up achieved by the fuel before discharge, and the time after discharge before reprocessing takes place. Significantly, though, from the point of view of deep geological disposal, variations in all or any of these factors would not materially affect the principal characteristics of the waste.

However, as a guide the Nuclear Energy Agency (NEA) of the Organization for Economic Cooperation and Development (OECD) [7] estimate about 5 m³ of conditioned HLW are produced per tonne of fuel processed, with 1 GW(e) of electrical power production accounting for about 6 m³ of solid vitrified HLW and about 50 m³ of other highly radioactive wastes characterized by high concentrations of α emitters. The total volume of HLW and other α-emitting wastes is therefore relatively small. Recent estimates suggest that the total nuclear capacity of the world could be as low as 600 GW(e) in the year 2000, which according to NEA estimates would result in about 15 000 tonnes of discharged spent fuel annually.

If only half this spent fuel can be processed at that time as a result of the limited reprocessing capacity available, it would result in about 750 m³ of solidified HLW [1]. Table 1.8 shows the amounts of waste arising from various fuel cycles per GW(a) of electricity generated. Evidence given to the United Kingdom House of Commons Environment Committee in March 1985 suggests that by the year 2000 about 4300 m³ of heat-generating wastes will have accumulated in the United Kingdom. This compares with 85 000 m³ of intermediate-level waste (ILW), also sometimes referred to as medium-level waste (MLW) and 650 000 m³ of low-level waste (LLW) over the same period. Table 1.9 gives the fuel-cycle wastes projected to have accumulated in the United States of America by the year 2000.

It should be mentioned that there is an argument put forward that reactor wastes should not be reprocessed as this significantly increases the volume of waste requiring disposal. From a volume of 4 m^{+3} arising from say one reactor over the course of one year's operation the amount of waste remaining after reprocessing is about 600 m^{+3} of LLW, 40 m^{+3} of ILW and a slightly reduced volume of 2.5 m^{+3} of HLW. This represents an overall increase in the volume of waste of about 160 times. Nevertheless nations continue to consider reprocessing important to recover depleted uranium for reuse in thermal reactors and for the recovery of plutonium for possible future use in fast breeder reactors.

1.4 THE NATURE OF HLW AND SURF

The main constituents of HLW and SURF are shown in Figs 1.3 and 1.4 respectively. In general SURF contains a greater actinide content than equivalent reprocessed HLW with typically up to 3% fission products as well as transuranic elements compared with 0.5% or less for reprocessed spent fuel. Table 1.10 lists the significant radionuclides most likely to be found in class I and II wastes. Table 1.11 shows how the level of radioactivity in SURF and HLW can be expected to evolve with time and it can be seen that SURF not only contains significantly greater radioactivity but remains active for considerably longer than reprocessed HLW. In general the activity is dominated for the first 300 years by the decay of cesium and strontium isotopes, which are β and γ emitters with half-lives of about 30 years. Between 300 and 3000 years americium-241, an α emitter, takes over, whilst longer-lived nuclides such as neptunium-237, an α emitter, iodine-129 and technetium-99, both of which are β emitters, remain active for millions of years.

Radioactive decay within both SURF and reprocessed HLW generates large quantities of heat as shown in Table 1.12. Again,

Table 1.8 Reprocessing waste for reference fuel cycles per GW(a) of electricity [8].

	LWR*		FBR†		HWR‡		HTR§
	1 Once-through	2 U–Pu cycle	3 U–Pu cycle	4 Once-through	5 U–Pu cycle	6 U–Th cycle	7 U–Th cycle
1. Hulls, spacers, insolubles							
Canister, HWR (kg)		33	86		31	49	—
Contained Pu (kg)		1·4	9·0		2·1	—	—
Radioactivity (1 year) (MCi)		1·4	1·6		0·5	0·3	0·3
2. Vitrified HLW							
Canister, HLWs (kg)		29	23		29	30	28
Contained Pu (kg)		2·3	15·1		3·6	11·9	6·2
Radioactivity (10 years) (MCi)		11·7	7·8		12·2	16·0	10·5
3. Noble gases							
Gas flasks (MCi)		17	17		17	18	18
Radioactivity (MCi)		0·3	0·2		0·3	0·8	0·6
4. Depleted uranium waste (as UO_2)							
Drums, unshielded (Mg)		13	1		83	—	—
Contained U		11·0	0·2		73	—	—

5. Medium-level and plant maintenance waste						
Drums, unshielded (kg)	54	35	—	118	190	3371
Drums, shielded	83	54	—	177	284	32
Contained Pu	0·9	5·9	—	1·5	—	0·3
Radioactivity (1 year) (kCi)	2·0	1·5	—	4·0	6·0	1·0
6. Low-level waste						
Drums, unshielded	113	74	—	244	392	44
Drums, shielded	13	8	—	27	43	5
Radioactivity (1 year) (kCi)	4	2·4	—	8	13	1·4
7. Plant decommissioning waste						
Drums, unshielded	140	91	—	302	484	54
Drums, shielded	16	10	—	33	54	6
Radioactivity (5 years) (kCi)	0·9	0·6	—	2	3	0·4

*LWR = light-water reactor.
†FBR = fast breeder reactor.
‡HWR = heavy-water-moderated reactor.
§HTA = high-temperature gas-cooled reactor.

Table 1.9   Fuel-cycle wastes projected for the US in the year 2000 (modified from [3].

	Total inventory		
Category of waste	Volume (10^3 m^3)	Activity (MCi)	Actinides (tons)
High-level solidified	10·5	126 000	1 110
Trans-uranium wastes			
Cladding hulls	10·3	587	75
Misc. α, β, γ solid	63·7	22	0·5
α solid	168·7	110	13
β–γ wastes			
Noble gases	0·48	1 170	
Iodine	0·01	0·006	
LWR tritium	586·1	2	
FP tritium	1·8	73	
Misc. β, γ solid	1 444·1	16	
Ore tailings	509 702·7	5·8	83 000

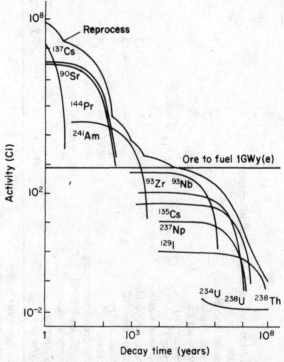

Figure 1.3   Total and individual activities of high-level waste from 1 Gwy(e) generation in a light-water reactor [9].

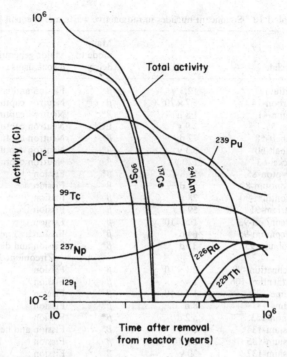

Figure 1.4 Radioactive decay of spent fuel from power water reactor. (Redrawn from [10].)

owing to the presence in the SURF of relatively larger quantities of unextracted uranium and plutonium it generates greater amounts of thermal power than the HLW resulting from spent-fuel reprocessing. In both cases, however, the thermal power decreases with the same characteristic half-lives as the radioactivity and will be reduced by a factor of 50 or more in a hundred years, while the contribution of the major fission products becomes insignificant within 200 to 250 years. Nevertheless the total amount of heat is considerable as shown in Fig. 1.5 which graphs the total heat released by buried SURF and HLW equivalent to 70 000 metric tonnes of uranium charged to a reactor [13]. Thermal anomalies associated with buried nuclear waste can be expected to last for up to 10 000 years or more. An idea of the enormous amounts of energy contained within HLW can be gained by examining the heat output in terms of equivalent barrels of oil. The annual fuel discharged from the Pickering A power station in Canada is estimated to equate to some 25 million barrels of oil [14].

It is also important to note that certain of the actinides found in SURF and reprocessed HLW are quite soluble in water as shown in Table 1.13. Whilst such factors as pH and E_h (redox potential) affect

Table 1.10 Significant nuclides in radioactive waste management [1].

Nuclide	Half-life	Major mode of decay	Major generation mechanisms
Tritium	12.3 y	β	Fission and neutron capture
Carbon-14	$5 \cdot 7 \times 10^3$ y	β	Neutron capture
Argon-41	1·8 h	$\beta*$	Neutron capture
Iron-55	2.9 y	EC†	Neutron capture
Cobalt-58	72 d	$\beta*$	Neutron capture
Cobalt-60	5·3 y	$\beta*$	Neutron capture
Nickel-63	120 y	β	Neutron capture
Krypton-85	10·8 y	$\beta*$	Fission
Strontium-89	51 d	β	Fission
Strontium-90	28 y	β	Fission
Yttrium-91	59 d	β	Fission
Zirconium-93	$1 \cdot 5 \times 10^6$ y	β	Fission
Zirconium-95	64 d	$\beta*$	Fission and neutron capture
Niobium-95	35 d	$\beta*$	Fission and daughter of Zirconium-95
Technetium-99	$2 \cdot 1 \times 10^5$ y	β	Fission
Ruthenium-106	1 y	β	Fission
Iodine-129	$1 \cdot 7 \times 10^7$ y	β	Fission
Iodine-131	8 d	$\beta*$	Fission
Xenon-133	5·2 d	$\beta*$	Fission
Cesium-134	2·1 y	$\beta*$	Fission and neutron capture
Cesium-135	2×10^6 y	β	Fission
Cesium-137	30 y	β	Fission
Cerium-141	33 d	$\beta*$	Fission
Cerium-144	285 d	$\beta*$	Fission
Promethium-147	2·6 y	β	Fission
Samarium-151	93 y	β	Fission
Europium-154	16 y	$\beta*$	Fission and neutron capture
Lead-210	22 y	β	Daughter of Polonium-214
Radon-222	3·8 d	α	Daughter of Radium-226
Radium-226	$1 \cdot 6 \times 10^3$ y	$\alpha*$	Daughter of Thorium-230
Thorium-229	$7 \cdot 3 \times 10^3$ y	$\alpha*$	Daughter of Uranium-233
Thorium-230	8×10^4 y	α	Daughter of Uranium-234
Uranium-234	$2 \cdot 4 \times 10^5$ y	α	Daughter of Protactinium-234
Uranium-235	$7 \cdot 1 \times 10^8$ y	$\alpha*$	Natural source, daughter of Plutonium-239
Uranium-238	$4 \cdot 5 \times 10^9$ y	α	Natural source
Neptunium-237	$2 \cdot 1 \times 10^6$ y	α	Neutron capture and daughter of Americium-241
Plutonium-238	87 y	α	Neutron capture and daughter of Curium-242
Plutonium-239	$2 \cdot 4 \times 10^4$ y	α	Neutron capture
Plutonium-240	$6 \cdot 6 \times 10^3$ y	α	Neutron capture
Plutonium-241	15 y	β	Neutron capture
Plutonium-242	$3 \cdot 87 \times 10^5$ y	α	Neutron capture
Americium-241	433 y	α	Neutron capture and daughter of Plutonium-241
Americium-243	$7 \cdot 37 \times 10^3$ y	α	Neutron capture
Curium-242	163 d	α	Neutron capture
Curium-244	18 y	α	Neutron capture

*With associated penetrating γ radiation.
†EC = orbital electron capture.

Table 1.11 Evolution of the radioactivity in spent fuel and high-level wastes per metric tonne of heavy metal in the original fuel element [11].

Time from reactor discharge (years)	Actinides and daughters in spent fuel	Actinides and daughters in high-level waste	Fission products in spent fuel or in high-level wastes
10	2660	120	11 500
100	234	32·7	1 270
1 000	57	8·3	0.8
10 000	16·3	0·94	—
100 000	1·4	—	—
1 000 000	0·5	—	—

Data illustrated are for a PWR with a fuel burn-up of 33 GW-day tonne^{-1} and subsequent reprocessing after 5 years, in TBq (1 TBq = 10^{12} disintegrations s^{-1}).

Table 1.12 Thermal power of spent fuel and high-level waste as a function of time in watts per metric tonne of heavy metal in the original fuel element [11].

Time from reactor discharge (years)	Spent fuel	High-level waste	Cladding hulls
10	1290	1120	33·5
100	284	134	1·46
1 000	49·4	6·8	0·25
100 000	1·0	0·10	—
1 000 000	0·3	0·10	—

Data illustrated are for a PWR with a fuel burn-up of 33 GW-day tonne^{-1} and subsequent reprocessing after 5 years. The thermal power of spent fuel from some other reactor types decreases more slowly.

the degree of solubility the potential leaching of actinides from class I and II wastes by groundwater and their subsequent transport back to the biosphere remains possible.

1.5 THE NEED FOR CONTAINMENT

The hazards of ionizing radiation to man have been evident ever since the discovery of X-rays in 1895 and it is now generally assumed that any exposure to ionizing radiations (i.e. γ and X-rays, neutrons, α and β particles) over and above natural background levels involves some degree of risk to health [6]. The main radiation types and their principal characteristics are listed in Table 1.14. The effects of

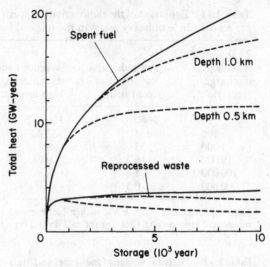

Figure 1.5 Heat released by buried wastes (solid curves) and heat remaining
in granite formation (dashed curves) for repositories at different
depths [12].

Table 1.13 Solubility of some
important actinides from high-
level reprocessing waste [15].

	Solubility $g/(1 \times 10^3)$ (p.p.m.)
SiO_2	120–20
^{238}U	485
^{237}Pu	2·4
$^{237}Np*$	0·007
^{230}Th	0·024
^{239}Pu	2·4

*Reducing conditions.

radiation are generally quantified in units of rads, which is shorthand
for radiation absorbed dose; one rad is the amount of radiation
required to liberate 1×10^{-5} J of energy per gram of absorbing
material. To help define the actual effects of such radiation on man a
second unit known as röntgen equivalent man or rem is used, with
one rem equal to one rad of X-rays or γ rays (see Table 1.6). It is
generally accepted that all radioactivity is potentially damaging to
living matter but its effects will depend, among other things, upon the
activity of the source, how close it is, how well shielded it is, the length
of time of exposure, the type of radiation and the tissue irradiated.

Table 1.14 Radiation types [2].

Radiation	Type	Charge	Energy	Average penetration	
				Air	Body
γ rays	Electromagnetic		0·1-40 MeV	No limit	Can pass through body
X-rays	Electromagnetic		10-100 keV	No limit	Penetrates deeply into body
α	Particle	+2	4-10 MeV	4-10 cm	Clothing, outer layers of skin
β⁻, β⁺	Particle	− or +	0·025-2·15 MeV	Several meters	Few mm into tissue
Neutrons	Particle	Zero	Several MeV	Captured by particles they collide with	

The levels of ionizing radiation associated with class I and II wastes could if unshielded severely affect human tissue, the gut and the central nervous system, resulting in leukopaenia, purpura haemor-rhaging, fever, vomiting and ultimately death, with lower doses contributing to cancers and genetic disorders. Table 1.15 shows some of the possible effects of whole-body radiation. To put the values given in Table 1.15 into perspective the annual whole-body dose rate due to natural radiation from rocks, cosmic rays and other sources is around 0.1 rem per year. Generally accepted man-made sources of radiation from medical uses of X-rays and colour television, for example, may increase the average total annual dose to about 0.2 rem per year. Table 1.16 shows the concentration of certain radioactive elements in various rocks and the estimated absorbed dose rates which they may give rise to. Table 1.17 shows how further background radiation can arise from the use of certain building materials. The level of exposure to natural background radiation will vary enormously from place to place according to the altitude, the nature of the rock, the ground and river water and many other factors, and it may not be appropriate to generalize on the extent of exposure in average terms. Nevertheless Table 1.18 gives a feel for the contribution that background radiation makes to the total exposure. Table 1.19 shows the relative contributions of a number of radiation sources to the population of the United Kingdom.

It is also a severe hazard to health for radionuclides to be ingested and the International Commission on Radiological Protection (ICRP) makes recommendations for the maximum concentrations of radionuclides permissible in air and water. The permissible levels of radionuclides recommended by the ICRP are extremely low, as can be seen from Table 1.20. This gives values in force in 1959 since when, if anything, the permitted levels have been reduced further [18]. In the United States Recommended Concentration Guide (RCG) values

Table 1.15 Effects of whole-body radiation [2].

Dose (rems)	Effects
0–25	A dose around 25 rem may reduce the white blood cell count
25–100	Nausea for about half those exposed, fatigue, changes to blood
100–200	Nausea, vomiting, fatigue, death possible, susceptible to infection (low white blood cell count)
200–400	A lethal dose for 50% of those exposed especially in absence of treatment. Bone marrow, spleen (blood-forming organs) damaged
>600	Fatal, probably even with treatment

Table 1.16 Typical concentrations of ^{40}K, ^{238}U and ^{232}Th in common rocks and soils, and the estimated absorbed dose rates in air at 1 m above the surface [16].

Type of rock or soil	Concentration Bq g^{-1} (pCi g^{-1})						Absorbed dose rate in air μGy h^{-1} (μrad h^{-1})	
	^{40}K		^{238}U		^{232}Th			
Igneous rock								
Granite	1.00	(27.0)	0.059	(1.60)	0.081	(2.20)	0.120	(12.0)
Diorite	0.70	(19.0)	0.023	(0.62)	0.033	(0.88)	0.062	(6.2)
Basalt	0.24	(6.5)	0.011	(0.31)	0.011	(0.30)	0.023	(2.3)
Dunite	0.15	(4.0)	0.0004	(0.01)	0.024	(0.66)	0.023	(2.3)
Sedimentary rock								
Limestone	0.09	(2.4)	0.028	(0.75)	0.007	(0.19)	0.020	(2.0)
Carbonate	—	—	0.027	(0.72)	0.008	(0.21)	0.017	(1.7)
Sandstone	0.37	(10.0)	0.019	(0.50)	0.011	(0.30)	0.032	(3.2)
Shale	0.70	(19.0)	0.044	(1.20)	0.044	(1.20)	0.079	(7.9)
Soil								
Serozem	0.67	(18.0)	0.031	(0.85)	0.048	(1.30)	0.074	(7.4)
Chernozem	0.41	(11.0)	0.021	(0.58)	0.036	(0.97)	0.051	(5.1)
Sodpodzolic	0.30	(8.1)	0.015	(0.41)	0.022	(0.60)	0.034	(3.4)
Boggy	0.09	(2.4)	0.006	(0.17)	0.006	(0.17)	0.011	(1.1)

Table 1.17 Concentrations of ^{40}K, ^{226}Ra and ^{232}Th in some building materials and the absorbed dose rate in air within a room constructed of them assuming 4π geometry [16].

Type of building material	Country	Concentration, Bq g^{-1} (pCi g^{-1})			Absorbed dose rate in air	
		^{40}K	^{226}Ra	^{232}Th	μGy h^{-1}	(μrad h^{-1})
Bricks	Sweden	0·93 (25)	0·096 (2·6)	0·126 (3·4)	0·33	(33)
	UK	0·63 (17)	0·052 (1·4)	0·044 (1·2)	0·16	(16)
	USSR	0·74 (20)	0·056 (1·5)	0·037 (1·0)	0·16	(16)
Concrete	Sweden	0·70 (19)	0·048 (1·3)	0·085 (2·3)	0·21	(21)
	UK	0·52 (14)	0·074 (2·0)	0·030 (0·8)	0·15	(15)
	USSR	0·56 (15)	0·033 (0·9)	0·030 (0·8)	0·12	(12)
Plaster	Sweden	0·02 (0·6)	0·003 (0·09)	<0·001 (<0·04)	<0·01	(<1)
	UK	0·15 (4·0)	0·022 (0·60)	0·007 (0·20)	0·04	(4)
	USSR	0·37 (10·0)	0·009 (0·25)	0·006 (0·17)	0·05	(5)
Granite	UK	1·04 (28)	0·089 (2·4)	0·085 (2·3)	0·28	(28)
	USSR	1·48 (40)	0·111 (3·0)	0·167 (4·5)	0·46	(46)
Rock aggregate	Sweden	0·81 (22)	0·048 (1·3)	0·070 (1·9)	0·20	(20)
	UK	0·81 (22)	0·052 (1·4)	0·004 (0·1)	0·12	(12)
Phosphogypsum	UK	0·07 (2)	0·777 (21·0)	0·019 (0·5)	0·68	(68)
	USA	—	1·480 (40·0)	0·007 (0·2)	1·26	(126)
Wood	Sweden	—	—	—	<0·004	(<0·4)

Table 1.18 Summary of estimated annual absorbed dose, from natural sources, to man living in areas of 'normal' radiation background [16].

Source of irradiation	Annual absorbed dose, μGy(mrad)			
	Gonads	Lung	Bone lining cells	Red bone marrow
External				
Cosmic rays:				
Ionizing component	280 (28)	280 (28)	280 (28)	280 (28)
Neutron component	3·5 (0·35)	3·5 (0·35)	3·5 (0·35)	3·5 (0·35)
Terrestrial radiation (γ)	320 (32)	320 (32)	320 (32)	320 (32)
Internal				
³H (β)	0·01 (0·001)	0·01 (0·001)	0·01 (0·001)	0·01 (0·001)
⁷Be (γ)	—	0·02 (0·002)	—	—
¹⁴C (β)	5·0 (0·5)	6·0 (0·6)	20 (2·0)	22 (2·2)
⁴⁰K (β+γ)	150 (15)	170 (17)	150 (15)	270 (27)
⁸⁷Rb (β)	8·0 (0·8)	4·0 (0·4)	9·0 (0·9)	4·0 (0·4)
²³⁸U–²³⁴U (α)	0·4 (0·04)	0·4 (0·04)	3·0 (0·3)	0·7 (0·07)
²³⁰Th (α)	0·04 (0·004)	0·4 (0·04)	8·0 (0·8)	0·5 (0·05)
²²⁶Ra–²¹⁴Po (α)	0·3 (0·03)	0·3 (0·03)	7·0 (0·7)	1·0 (0·1)
²¹⁰Pb–²¹⁰Po (α+β)	6·0 (0·6)	3·0 (0·3)	34 (3·4)	9·0 (0·9)
²²²Rn–²¹⁴Po (α)inhalation	2·0 (0·2)	300 (30)	3·0 (0·3)	3·0 (0·3)
²³²Th (α)	0·04 (0·004)	0·4 (0·04)	7·0 (0·7)	0·4 (0·04)
²²⁸Ra–²⁰⁸Tl (α)	0·6 (0·06)	0·6 (0·06)	11 (1·1)	2·0 (0·2)
²²⁰Rn–²⁰⁸Tl (α)inhalation	0·08 (0·008)	40 (4)	1·0 (0·1)	1·0 (0·1)
TOTAL	780 (78)	1100 (110)	860 (86)	920 (92)

Table 1.19 Comparison with other sources of exposure [17].

Source	Annual collective effective dose equivalent to the UK population (man Sv)
Natural background*	104 000
Medical irradiation	28 000
Fallout	560
Miscellaneous sources	450
Occupational exposure	500
Total	134 000
Effluents from the civil nuclear programme in 1978	160†

*Including exposure to radon daughters.
†Collective dose commitment equated with the annual collective dose.

are used in a similar manner and on occasion the toxicity of a nuclear waste may be given in terms of the amount of water required to dilute the waste to the point where it complies with the relevant RCG. The reader may come across the deleterious effects of radiation expressed in terms of 'stochastic' and 'non-stochastic' effects. Stochastic effects are those for which the probability of an effect occurring, rather than its severity, is regarded as a function of dose, without threshold. Non-stochastic effects are those for which a threshold may therefore occur.

Clearly, then, SURF, reprocessed HLW and other α wastes need to be contained in an environment which provides adequate shielding and isolation from man and the biosphere generally, whilst at the same time being capable of absorbing and dispersing the thermal power generated by the radioactive decay of the waste. The exact levels of radiation which are considered safe is a matter of some controversy. Recommended safety levels are constantly being updated and have generally been reduced from year to year. Pentreath [6] gives a detailed discussion of the issues and principles involved in determining and setting limits of safe exposure to the effects of ionizing radiation and the ingestion of radioactive material. The underlying principle on which ICRP recommendations are based is that risks to individuals should be kept to acceptable levels. In order to achieve this objective, ICRP recommends a system of dose limitation which has three main requirements:

1. 'that no practice shall be adopted unless its introduction produces a positive net benefit' (justification);
2. 'that all exposures shall be kept as low as reasonably achievable, economic and social factors being taken into account' (also known as optimization of protection and the ALARA principle);

3. 'the dose equivalent to individuals shall not exceed the limits recommended for the appropriate circumstances by the Commission' (dose limits).

These principles have been applied to some solid-waste-disposal management practices but it is recognized that there may be some difficulties in applying the ICRP system of dose limitation to practices involving long-lived radionuclides [19].

In 1984 a group of NEA experts recommended that the appropriate form for a limit on individual detriment from waste disposal would be in the form of risk, where risk is defined as the product of the probability of incurring a radiation dose and the probability of that dose giving rise to deleterious health effects. The risk limits recommended are generally comparable with the current ICRP annual dose limits. The United Kingdom Department of the Environment has produced guidelines for the establishment of a HLW repository [21]. These guidelines state:

'The risk to any member of the public in any one year from exposure to radiation from all sources other than background and medical exposure, should not be greater than that associated with a dose of 1 mSv. (In the United Kingdom, the assumed annual risk of death from radiation exposure of 1 mSv per year is 1 in 100 000.) In order to take account of exposure pathways and health effects not at present recognised, and avoid prejudicing any future decisions that might lead to other activities which cause radiation exposure to the same members of the public, an additional margin of safety is required. *The appropriate target applicable to a single repository at any time is, therefore, a risk to an individual in a year equivalent to that associated with a dose of 0.1 mSv: about 1 chance in a million. The risk from this dose would be substantially less than those arising from variations in natural background radiation within the United Kingdom.'*

1.6 THE CONCEPT OF GEOLOGICAL DISPOSAL OF RADIOACTIVE WASTES

It is possible to store highly radioactive waste at the Earth's surface but the time scales involved before the waste becomes safe are considerable when compared with the periods of freedom from war and natural disaster that have historically occurred. Deep burial in impermeable rock in areas known to be geologically stable has certain attractions in contrast to the surface storage option. The time periods involved before stored waste becomes safe will be extensive. While most radionuclides are unlikely to pose a threat for much more than 1000 years the major isotopes of plutonium and americium will remain active for 100 000 years and a few nuclides such as iodine-129

Table 1.20 Maximum permissible concentrations of some selected radionuclides in air and in water for occupational exposure (1 μCi = 3·7 × 10⁴ Bq) (modified from [18]).

Radionuclide and type of decay		Critical organ	Maximum permissible concentrations (μCi cm^{-3})			
			For 40-h week		For 168-h week	
			$(MPC)_w$	$(MPC)_a$	$(MPC)_w$	$(MPC)_a$
^{3}H (β^-)	sol.	Body tissue	0·1	5×10^{-6}	0·03	2×10^{-6}
^{14}C (β^-)CO_2	sol.	Fat	0·02	4×10^{-6}	8×10^{-3}	10^{-6}
^{32}P (β^-)	sol.	Bone	5×10^{-4}	7×10^{-8}	2×10^{-4}	2×10^{-8}
	insol.	Lung		8×10^{-8}		3×10^{-8}
^{35}S (β^-)	sol.	GI(LLI)*	7×10^{-4}	10^{-7}	2×10^{-4}	4×10^{-8}
	sol.	Testis	2×10^{-3}	3×10^{-7}	6×10^{-4}	9×10^{-8}
	insol.	Lung		3×10^{-7}		9×10^{-8}
^{41}Ar (β^-, γ)		GI(LLI)	8×10^{-3}	10^{-6}	3×10^{-3}	5×10^{-7}
^{54}Mn (γ)	sol.	Total body	4×10^{-3}	2×10^{-6}	10^{-3}	4×10^{-7}
		GI(LLI)	0·01	8×10^{-7}		3×10^{-7}
	insol.	Liver		4×10^{-7}	4×10^{-3}	10^{-7}
		Lung		4×10^{-7}		10^{-8}
^{59}Fe (β^-, γ)	sol.	GI(LLI)	3×10^{-3}	6×10^{-7}	10^{-3}	2×10^{-7}
	sol.	GI(LLI)	2×10^{-3}	4×10^{-7}	6×10^{-4}	10^{-7}
		Spleen	4×10^{-3}	10^{-7}	10^{-3}	5×10^{-8}
	insol.	Lung		5×10^{-8}		2×10^{-8}
^{60}Co (β^-, γ)	sol.	GI(LLI)	2×10^{-3}	3×10^{-7}	5×10^{-4}	9×10^{-8}
	sol.	GI(LLI)	10^{-3}	3×10^{-7}	5×10^{-4}	10^{-7}
	insol.	Lung		9×10^{-9}		3×10^{-9}
	sol.	GI(LLI)	10^{-3}	2×10^{-7}	3×10^{-4}	6×10^{-8}
^{65}Zn (β^+, γ)	sol.	Total body	3×10^{-3}	10^{-7}	10^{-3}	4×10^{-8}
		Prostate	4×10^{-3}	10^{-7}	10^{-3}	4×10^{-8}
		Liver	4×10^{-3}	10^{-7}	10^{-3}	5×10^{-8}
		Lung		6×10^{-8}		2×10^{-8}
	insol.	GI(LLI)	5×10^{-3}	9×10^{-7}	2×10^{-3}	3×10^{-7}

^{85}Kr (β^-)		Total body		10^{-5}		3×10^{-6}
^{90}Sr (β^-)	sol.	Bone	10^{-5}	10^{-9}	4×10^{-6}	4×10^{-10}
	insol.	Lung		5×10^{-9}		2×10^{-9}
		GI(LLI)	10^{-3}	2×10^{-7}	4×10^{-4}	6×10^{-8}
^{95}Zr (β^-, γ)	sol.	GI(LLI)	2×10^{-3}	4×10^{-7}	6×10^{-4}	10^{-7}
		Total body	3	10^{-7}	1	4×10^{-8}
	insol.	Lung		3×10^{-8}		10^{-8}
		GI(LLI)	2×10^{-3}	3×10^{-7}	6×10^{-4}	10^{-7}
^{95}Nb (β^-, γ)	sol.	GI(LLI)	3×10^{-3}	6×10^{-7}	10^{-3}	2×10^{-7}
		Total body	10	5×10^{-7}	4	2×10^{-7}
	insol.	Lung		10^{-7}		3×10^{-8}
		GI(LLI)	3×10^{-3}	5×10^{-7}	10^{-3}	2×10^{-7}
^{99}Tc (β^-)	sol.	GI(LLI)	$0{\cdot}01$	2×10^{-6}	3×10^{-3}	7×10^{-7}
	insol.	Lung		6×10^{-8}		2×10^{-8}
		GI(LLI)	5×10^{-3}	8×10^{-7}	2×10^{-3}	3×10^{-7}
^{106}Ru (β^-, γ)	sol.	GI(LLI)	4×10^{-4}	8×10^{-8}	10^{-4}	3×10^{-8}
	insol.	Lung		6×10^{-9}		2×10^{-9}
		GI(LLI)	3×10^{-4}	6×10^{-8}	10^{-4}	2×10^{-8}
^{129}I (β^-, γ)	sol.	Thyroid	10^{-5}	2×10^{-9}	4×10^{-6}	6×10^{-10}
	insol.	Lung		7×10^{-8}		2×10^{-8}
		GI(LLI)	6×10^{-3}	10^{-6}	2×10^{-3}	4×10^{-7}
^{131}I (β^-, γ)	sol.	Thyroid	6×10^{-5}	9×10^{-9}	2×10^{-5}	3×10^{-9}
	insol.	Lung		3×10^{-7}		10^{-7}
		GI(LLI)	2×10^{-3}	3×10^{-7}	6×10^{-4}	10^{-7}
^{133}Xe (γ)		Total body		10^{-5}		3×10^{-6}
^{137}Cs (β^-, γ)	sol.	Total body	4×10^{-4}	6×10^{-8}	2×10^{-4}	2×10^{-8}
		Liver	5×10^{-4}	8×10^{-8}	2×10^{-4}	3×10^{-8}
		Spleen	6×10^{-4}	9×10^{-8}	2×10^{-4}	3×10^{-8}
		Muscle	7×10^{-4}	10^{-7}	2×10^{-4}	4×10^{-8}
	insol.	Lung		10^{-8}		5×10^{-9}
		GI(LLI)	10^{-3}	2×10^{-7}	4×10^{-4}	8×10^{-8}

*GI(LLI) is the lower large intestine.

will remain active for several million years. Whilst extensive in human terms such time periods are mere moments in the geological sense. Gera [22] has argued that no geological disposal system can be expected to ensure absolute containment of buried HLW and suggests it is certain that eventually some radionuclides from buried waste will return to the biosphere. He goes on, however, to make the point that it would be wrong to believe geological disposal was therefore unacceptable as the dose equivalent rate reaching the biosphere would with proper repository design be very low. After as little as 100 000 years the dose equivalent rate of buried HLW would be much the same as that resulting from naturally occurring uranium deposits. On this basis 100 000 years is generally taken as being a minimum period of time for confinement of waste within a geological repository. According to the NEA [1] this conclusion is based mainly on the fact that the radiotoxicity for ingestion of the waste in a 100 000-year-old repository and of a large pile of uranium tailings are very similar. This assumes of course that the effects of uranium ore tailings are themselves acceptable. However, in this context there is the obvious and important difference that in geological disposal schemes the waste would be buried at some depth while ore tailings are at the surface with ready access to the biosphere.

Geological disposal has the advantage that it can be designed to be an entirely passive system with no long-term monitoring requirement. Deep burial should ensure that little or no radioactivity escapes at the Earth's surface for a predetermined time period, while adequate site selection with respect to stability and groundwater flow regimes should ensure that waste once placed underground remains there for the most part. Deep burial also reduces the risk of inadvertent disturbance by man and deters any unauthorized interference with or removal of the waste. Equally important, geological disposal could be carried out using existing proven mining, engineering and deep drilling techniques.

There are two principal philosophies of waste disposal by burial which have been applied to domestic and industrial waste, and which according to the British Geological Survey in evidence given to the House of Commons Environment Committee on 19 June 1985 are equally appropriate to radioactive wastes: (1) Containment, whereby the waste remains always within the confines of the disposal facility. (2) Dilute and disperse, whereby components of the waste migrate away from the site in a controlled or predictable fashion in mobile groundwaters. The objective is that the toxic waste be dispersed throughout a large volume of rock at low concentrations.

The latter approach is generally accepted to have greater advantages for the following reasons: (1) Containment is difficult to ensure over long periods, relying completely on the integrity of engineered barriers such as the waste matrix and its container. For long-lived radioactive wastes it is in any case impossible to assure containment for the time periods of concern. (2) A containment philosophy means that the waste is always present at high concentrations and consequently constitutes a long-term hazard at one site. Processes of dispersion and degradation can reduce this hazard substantially.

Nevertheless the necessary isolation of HLW within a geological environment would not necessarily be achieved by the nature of the rock alone which would, according to most currently accepted ideas, form only one part of a multi-barrier system. Solidification of the waste by a process of vitrification or fixing it within an artificial rock medium can reduce its rate of leaching should groundwater gain entry to the repository, whilst suitable waste containers and buffer materials can further reduce the mobility of the waste. Finally, such waste as does escape the repository will be subjected to a series of retention and dilution mechanisms as it passes through the geosphere before emerging in the biosphere. Figure 1.6 shows the toxic potential of a typical HLW.

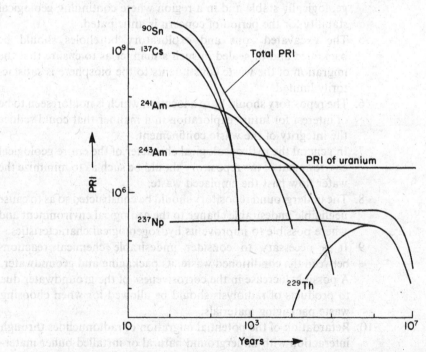

Figure 1.6 Typical toxic potential (PRI] of vitrified high-level waste from reprocessing one tonne of LWR fuel as a function of time compared with the toxic potential of an equivalent amount of a 0.17% uranium ore [23].

1.7 CRITERIA FOR A HLW GEOLOGICAL REPOSITORY

The IAEA has produced a list of twelve criteria to be taken into account in the implementation of underground nuclear waste disposal systems [4]:

1. The geology of the area should be investigated beyond the repository zone so as to permit an effective evaluation of performance.
2. The wastes should be emplaced deep enough so that they will not be laid open by natural processes (e.g. by erosion or uplift) while their radioactivity level is unacceptably high. Possible future climatological changes should be considered.
3. The geological medium in the zone of waste emplacement should be sufficiently thick and extend far enough laterally to provide an adequate protective zone around the repository. Such a zone would assist in providing an adequate degree of separation from underlying and overlying strata and flanking transition zones and faults, etc.
4. The geological host medium and its surroundings should be geologically stable and in a region where continuing geological stability for the period of concern is anticipated.
5. The excavated zone and exploratory boreholes should be backfilled and/or sealed in such a manner as to ensure that the migration of the waste constituents to the biosphere is satisfactorily limited.
6. The repository should be in a location which is not foreseen to be of interest for future exploration in a manner that could reduce the integrity of the waste confinement.
7. In general the hydrogeological character of the entire geological environment of the repository should be such as to minimize the water flow past the emplaced waste.
8. The underground repository should be constructed so as to cause negligible undesirable change to the geological environment and where possible to improve its hydrogeological characteristics.
9. It is necessary to consider undesirable chemical reactions between the conditioned waste, its packaging and groundwater. A possible increase in the corrosiveness of the groundwater due to products of radiolysis should be allowed for when choosing waste packaging materials.
10. Retardation of the potential migration of radionuclides through interaction with underground natural or installed buffer materials should be considered.
11. The engineered structures and geological strata should be capable of withstanding the effects of the radiation and heat generated by the wastes.

12. The repository should be designed so as to minimize the possible creation of waste migration pathways by failure of engineered structures.

As well as working through these twelve criteria when designing a HLW repository it helps conceptually to think of a repository as consisting of three pillars as identified by Zurkinden and Niederer [24]: the repository, the waste packaging, and the waste itself as shown in Fig. 1.7, each part being designed to keep the waste in and the groundwater out for a sufficiently long time to ensure that none of the waste finds its way back to the biosphere for 100 000 years or longer.

The potential of different rock types as HLW repositories will vary according to their mineralogy, their geological stability, their mechanical stability, particularly during heating and subsequent cooling, and their hydrogeology. Most rock types are of interest as HLW repositories including representatives from the sedimentary, igneous and metamorphic groups. Table 1.21 gives the average composition for the main rock types. Of particular interest are the evaporites, clays and granites, but research is also taking place into the potential of gabbros, basalts, tuffs and diabases. The extensive range of rock types involved and the extent of international involvement in the problem of HLW disposal can be gauged from Table 1.22. With some 35 countries now generating nuclear power

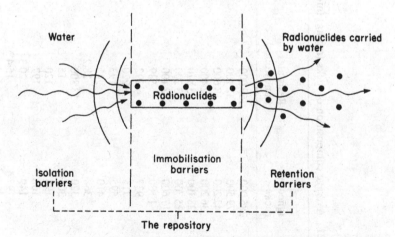

Figure 1.7 Groundwater transport and barriers, 'the three-pillar strategy'. Isolation barriers include the host rock, rock-cavity lining and corrosion-resistant containers. Immobilization barriers are mainly based on fixing the waste in a leach-resistant solid matrix of glass or synthetic rock. Retention barriers include the host rock, backfilling material and surrounding geological media. (Modified from [24].)

Table 1.21 Average composition of rocks (mg/kg) (modified from [25]).

	Igneous rocks	Resistates (sandstones)	Hydrolysates (argillaceous rocks)	Precipitates (carbonate rocks)	Evaporites
Si	285 000	359 000	260 000	33 900	386
Al	79 500	32 100	80 100	8 970	29
Fe	42 200	18 600	38 800	8 190	265
Ca	36 200	22 400	22 500	272 000	11 100
Na	28 100	3 870	4 850	393	310 000
K	25 700	13 200	24 900	2 390	4 280
Mg	17 600	8 100	16 400	45 300	3 070
P	1 100	539	733	281	
Mn	937	392	575	842	4.4
F	715	220	560	112	24
Ba	595	193	250	30.1	173
S	410	945	1 850	4 550	26 800
Sr	368	28.2	290	617	234
C	320	13 800	15 300	113 500	525 000
Cl	305	15	170	305	
Cr	198	120	423	7·08	10·6
Rb	166	197	243	46·0	
V	149	20·3	101	12·6	0·3
Cu	97·4	15·4	44·7	4·44	2

Ni	93·8	2·57	29·4	12·8	1·4
Zn	80	16·3	130	15·6	0·6
N	46		600		
Li	32·2	15	46·2	5·16	30
Co	23	0·328	8·06	0·123	1·6
Ga	18·5	5·87	22·8	2·69	
Pb	15·6	13·5	80	16·5	0·9
Th	11·4	3·94	13·1	2·01	0·2
B	7·5	90	194	16	1·2
Cs	4·3	2·15	6·2	0·771	
Be	3·65	0·258	2·13	0·175	0·2
U	2·75	1·01	4·49	2·2	
Sn	2·49	0·115	4·12	0·166	
Br	2·37	1·0	4·3	6·6	33
As	1·75	1·0	9·0	1·75	
W	1·42	1·56	1·92	0·561	
Ge	1·39	0·881	1·32	0·363	
Mo	1·25	0·50	4·25	0·75	1·5
I	0·45	3·75	4·4	1·59	1
Hg	0·328	0·0574	0·272	0·0456	
Cd	0·192	0·0199	0·183	0·0476	
Ag	0·151	0·122	0·271	0·189	
Se	0·050	0·525	0·60·	0·315	0·2
Au	0·00357	0·00457	0·00345	0·00179	

Table 1.22 Survey of best facilities for HLW disposal [7].

						Test facilities at depth	
	Country	Generic studies	Survey of sites	Surface investigations	Test boreholes	Access shaft or tunnel sunk	Test facility*	Location
Crystalline rocks	USA	×	×					
	Switzerland	×	×	×				
	Sweden	×	×	×				
Granite	Finland	×	×	×				
	Canada	×	×	×	×	×		Whiteshell
	France	×	×	×	×			
	Japan	×	×					
	Spain	×	×					
	Sweden	×	×	×	×		×	Stripa
	Switzerland	×	×	×	×	×	×	Grimsel Pass
	UK	×	×	×	×			
	USA	×	×	×	×	×	×	Climax
Other								
Gabbro	Sweden	×		×	×			
	Canada			×	×			
Diabase	Japan			×	×			
Evaporites								
Salt Diapirs	Denmark		×	×	×			
	F.R. Germany	×	×	×	×	×		Gorleben†
	F.R. Germany	×	×	×	×		×	Asse II†‡§
	Netherlands	×	×	×				

	Avery Island	Lyons	WIPP†	Felsenau	Mol†	Konrad††	NSIF Hanford†	NIS†
Bedded salt								
USA	×				×			
Spain		×			×			
USA		×			×			
USA		×			×			
Switzerland								
USA								
Anhydrite								
Other sedimentary rocks								
Clay								
Belgium	×	×	×	×	×	×		
Italy								
Switzerland								
UK	×	×	×	×	×	×	×	
USA	×	×	×	×	×	×	×	
Japan								
Spain								
Shale								
Other								
Mixed marine sedimentary sequence								
F.R. Germany						×	×	
Basalt								
USA								
USA								
Japan								

*Test facilities at depth without a shaft being sunk implies the facility is in an existing mine.
†This facility or another facility at this site has been accepted or is planned to accept waste.
‡Low and intermediate waste only; no high-level waste.
§The Asse mine is situated in a salt anticline.

and producing their own nuclear wastes, the number of research and development programmes in the field of HLW disposal has increased steadily. At the same time there has developed a high degree of international cooperation on the matter. The United States of America, for instance, has bilateral research agreements with seven other countries while the NEA is more and more taking on a worldwide coordinating role.

Ideally rock to be used as a geological repository will have the following characteristics [7]:

1. Hydrogeological properties which minimize the extent to which waste is exposed to moving groundwater.
2. Geochemical and mineralogical properties which will ensure that migrating radionuclides will be retarded or immobilized before reaching the biosphere.
3. Thermochemical properties which can allow for heat loading due to waste emplacement without damaging the structural competency of the repository or host medium.
4. Adequate structural strength and sufficient stability to ensure that the physical integrity of a repository is not jeopardized during the operational period.

The multi-barrier concept in its purest philosophical form would require that all four of these attributes should be present at any individual site. It may be that certain failures to meet these criteria could be overcome by engineered barriers.

While the emphasis until now has been placed upon the clays, shales, evaporites and igneous rocks, Chapman and McEwen [26] have recently argued that other rock types, including massive sandstones, for example, may also be suitable provided their hydrogeological setting is appropriate. Fig. 1.8 shows the locations and basic geology of areas under investigation as potential HLW repositories in the United States of America. The rock types involved include bedded salt, domed salt, granites, basalts and tuffs and argillaceous rocks. In the United Kingdom some 127 areas were originally identified as being potentially capable of meeting the site selection criteria [27]. Of the 127 sites 104 contain igneous or metamorphic rocks, 17 argillaceous sediments, and six bedded evaporite sequences. While domed salts are present beneath the North Sea they are not well known beneath the United Kingdom itself, and unlike Belgium and the United States of America, for example, the United Kingdom has no major research effort into domed salts as potential repositories. The main areas of initial interest in the United Kingdom are shown in Fig. 1.9.

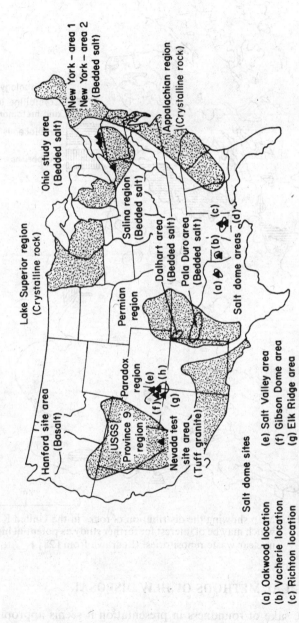

Figure 1.8 Areas under investigation for nuclear waste disposal in the United States of America. (Modified from [10].)

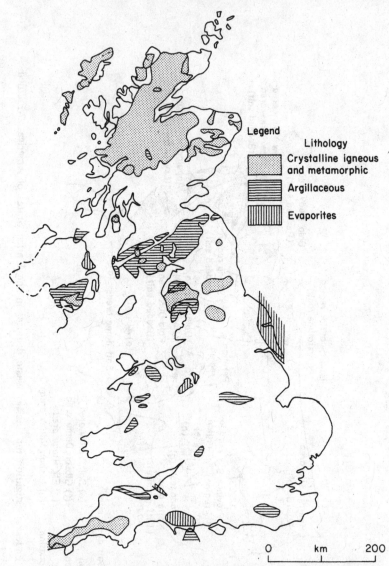

Figure 1.9 Map showing the distribution of rocks in the United Kingdom
which may be of interest for further study as potential high-level
nuclear waste repositories. (Redrawn from [28].)

1.8 NON-GEOLOGICAL METHODS OF HLW DISPOSAL

For the sake of roundness in presentation it seems appropriate to
place the geological disposal option in context by briefly mentioning
alternative disposal methods which have been or are being considered. An initially attractive idea was to place waste in space so

removing it from the Earth altogether. Besides being very expensive such a scheme, to be effectively safe, would require a reasonable certainty that the launch vehicle would achieve its designed orbit. As demonstrated by the many launch-vehicle accidents of 1985 and 1986, including the total destruction of Ariane and Titan rockets together with the tragic loss of *Challenger*, space flight is not yet able to put HLW safely into the Sun or outside the Earth's immediate environment.

Zeller, Saunders and Angino [27] have suggested that the ice caps could provide a suitable resting place for HLW. The main advantages of the ice caps are their remoteness and the simplicity of only having to place the waste on the ice surface to dispose of it. The heat generated by the decaying waste would melt the ice and allow the waste in its container to sink to the bottom of the ice caps. Assuming an average ice thickness of 3 km a canister would take no more than ten years to reach bedrock. The problems with such a disposal concept relate mainly to the often transient nature of ice and its dependence for existence upon a climate that we know can change in tens of thousands of years, much shorter than the hundreds of thousands of years of isolation required. There may also be problems with areas of unfrozen water occurring within the ice.

The idea of transmuting the waste has also been proposed. This involves the transmutation of the long-lived actinide radionuclides within a reactor, the main aim being to remove the troublesome actinide waste from the HLW and reintroduce it into a reactor core. Inside the reactor core different nuclides would be created by neutron capture. Some of these new nuclides would subsequently undergo fission giving rise to shorter-lived and more easily disposable fission-product radionuclides. However, this process can also produce heavier, more hazardous radionuclides by neutron capture. Clearly a balance needs to be struck and appropriate ratios calculated. For the moment the idea is not capable of coping with the quantities of waste requiring disposal, but the advent of fast breeder reactors may further research in this field.

1.9 CONCLUSIONS

Considerable quantities of HLW already exist and are awaiting disposal. It is the nature of HLW that it is highly radioactive and may also generate large quantities of heat as it decays. HLW requires isolation from the biosphere for between 100 000 and 1 000 000 years to comply with the available health and 'acceptable risk' guidelines. While such waste can be contained in surface storage cooling tanks such a measure is generally regarded as only an interim arrangement.

2   *The suitability of evaporites as HLW repositories*

2.1 INTRODUCTION

Evaporites have long been considered as a prime prospect for HLW disposal [29]. This is due to their low water content, high thermal conductivity and ability to act in a plastic manner which can make them self-sealing should fractures form. Evaporites, which occur in all the continents, are mostly formed from minerals which precipitated under hot arid climatic conditions in seas and large saline lakes. The resulting beds of evaporite minerals can be very thick and extensive. The Permian salts of the Delaware Basin for instance extend over 300 000 km^2, while in Texas salt deposits up to 3500 m thick have been proven. Most of the evaporite deposits under investigation for nuclear waste disposal are, however, much thinner than this though still substantial, such as the 1200 m thick sequence being investigated as part of the Waste Isolation Pilot Plant (WIPP) Project near Carlsbad in New Mexico [30].

2.2 MINERALOGY AND VARIABILITY OF EVAPORITES

Evaporite deposits are extremely variable in both their mineralogy and their stratigraphy. The main minerals and later replacement products of evaporites are given in Table 2.1. About 120 minerals have been identified in marine and non-marine salt deposits but many of these are inherently unstable and do not usually occur in older evaporite rock. In general only some forty or so minerals commonly occur in evaporites and these can be conveniently subdivided on the basis of their anion content as carbonates, sulphates, chlorides and borates [31]. Most evaporites have a seawater origin and the predominance of chlorides (94.5% of anion molarity of seawater) and sulphates (4.9% of anion molarity of seawater) in seawater is reflected in the predominance of halite, gypsum and anhydrite deposits in evaporite sequences. Non-marine evaporites also occur as playa and saline lake deposits and these may often be distinguished by the occurrence of minerals such as trona, mirabilite and glauberite which are rarely found in marine evaporite deposits.

Owing to their high solubility evaporites are only rarely seen at

Table 2.1 Minerals occurring as primary constituents or later replacement products in British evaporites [31].

Predominant constituents	Minor constituents
Dolomite ($CaMgCO_3$)	Magnesite ($MgCO_3$)
Halite (NaCl)	Hematite (Fe_2O_3)
Polyhalite ($K_2MgCa_2^*(SO_4)_4 \cdot 2H_2O$)	Kieserite ($MgSO_4 \cdot H_2O$)
Gypsum ($CaSO_4 \cdot 2H_2O$)	Carnallite ($MgKCl_3 \cdot 6H_2O$)
Anhydrite ($CaSO_4$)	Quartz and clay minerals,
Sylvine (KCl)	some sulphides

outcrop except in arid areas. Where they are seen at the surface it is normally as gypsum and not as halite which is some 150 times more soluble. At depth, however, halite is more common than gypsum, reflecting the relative abundance of sodium and chloride ions in seawater.

Usiglio [32] showed in 1849 that salts are precipitated from seawater in an ordered sequence. When the original volume of seawater is reduced by evaporation to about one-half a little iron oxide and some $CaCO_3$ are precipitated. When the volume has been reduced further to about one-fifth gypsum is formed. Upon reduction to about one-tenth of the original seawater volume, NaCl begins to crystallize. Reduction in water volume below this leads to the appearance of sulphates and chlorides of magnesium and finally to NaBr and KCl. Owing to alterations in sea level and climatic changes the extent to which individual evaporite cycles will be completed varies considerably. Additionally, between periods of salt formation normal sedimentary processes may resume, bringing in large amounts of clay and quantities of sand, or extensive thicknesses of limestone may develop. The result is that it is not uncommon to find evaporites interbedded with considerable thicknesses of shale, sandstone and limestone, all with markedly different hydrogeological, physical and chemical properties from their neighbouring evaporites. Brookins [10] has shown a 'typical generic' bedded evaporite sequence based on a range of actual evaporite logs and this is reproduced in Fig. 2.1.

Evaporite minerals are very susceptible to alteration following their initial precipitation. Gypsum can be altered to anhydrite when buried at depth, where the increase in both the pressure and temperature combine to drive off the two molecules of water in the gypsum. Massive anhydrite tends to be impermeable and chemically stable and in Switzerland is under investigation as a suitable medium

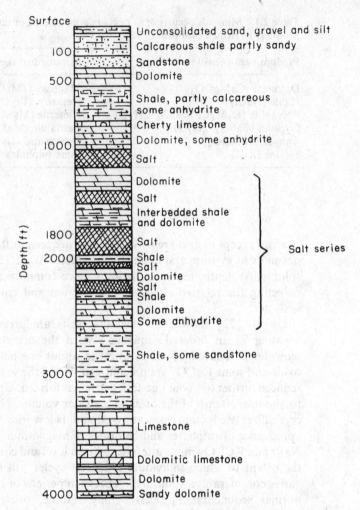

Figure 2.1 Generic stratigraphic section of a typical bedded salt sequence [10].

for low-level waste disposal. Availability of groundwater can, however, reverse anhydrite to gypsum with an associated increase in volume which could potentially make anhydrite repositories self-sealing if groundwater was able to enter. Against this it is necessary to recognize that anhydrite has a tendency to be brittle and is prone to fracturing which may increase the chances of groundwater reaching the waste in the first place and pose problems in repository construction.

While monomineralic evaporite units of adequate thickness, about 75 m [10], commonly occur, and it is these which are of interest as HLW repositories, it is important to be aware of the variability which can occur in evaporite sequences as shown in Table 2.2.

Table 2.2 Analysis of selected evaporite units [33].

	1	2	3		4	5	6
Al_2O_3 } Fe_2O_3	0·10	—	—	Mg	0·08	0·34	7·09
				Ca	0·15	4·15	0·59
CaO	32·37	40·61	18·71	Na	35·55	28·09	6·69
				K	4·58	6·44	11·40
				SO_4	0·36	11·16	1·41
MgO	—	—	6·72	Cl	59·20	49·11	41·31
K_2O } Na_2O	0·10	—	—	H_2O	0·13	0·32	31·51
		—	15·64	SiO_2	—	0·40	—
H_2O	20·94	1·87	6·08	Al_2O_3	—	0·17	—
SO_3	46·18	56·82	53·04				

1. Gypsum rock, Hillsborough, New Brunswick, Canada.
2. Anhydrite rock, Gypsum, Colorado, USA.
3. Polyhalite rock, Aislaby, Yorkshire, UK.
4. Halite–sylvite rock, Nova Scotia.
5. Halite–anhydrite–sylvite rock, Gluckauf-Sonderhausen, Germany.
6. Carnallite–halite rock, Germany.

Original evaporite sequences are stratified or bedded, but when loaded by later sedimentary overburden the salts, being less dense than the overlying sediments and capable of plastic deformation, are forced upwards as salt diapirs. This diapiric movement can result in the bending of overlying strata to form anticlinal structures. The columns of salt, which can be a few kilometres across and several kilometres high [34], can penetrate through the overlying rock and rise to very near the surface where exceptionally they may actually outcrop. This phenomenon is due to the unusual physical properties of certain evaporites, most notably rock salt. The density of evaporites is anomalously low as shown in Fig. 2.2. While most other sediments originally have lower densities than salt this situation reverses as these sediments undergo diagenesis, progressively becoming more dehydrated, compacted and cemented. Salt, on the other hand, remains almost incompressible throughout subsequent burial and metamorphism. In comparison with most rocks, salt then becomes relatively less dense, and this combined with its ability to flow often results in its early migration. The tops of salt domes are found at a depth of only 900 m in the Gulf Interior Basin of the United States of America, for instance, where they are under investigation as

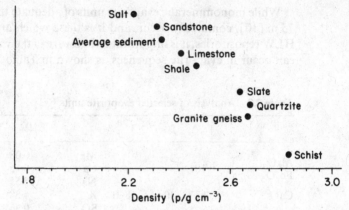

Figure 2.2   Density of salt in comparison with other rock types. (Modified from [35].)

potential disposal sites. Amongst other places salt diapirs are being examined at Gorleben in the Federal Republic of Germany [37] and in the Mors area of Denmark [38]. Migrating salt can also take on a number of forms besides the classic salt dome shape and these are summarized in Fig. 2.3. Caution is therefore needed in assessing and describing the shape of diapiric salt structures bearing in mind the enormous diversity of forms found within the geological column.

2.3   PHYSICAL PROPERTIES OF EVAPORITES

The main physical properties of evaporitic rocks ·are given in Table 2.3 [39]. In the rock strength classification given by the Geological Society of London [40] anhydrite is a strong rock,

Table 2.3   Some physical properties of evaporitic rocks [40].

	Gypsum	Anhydrite	Rock salt	Potash
Relative density	2·36	2·93	2·2	2·05
Dry density ($Mg\,m^{-3}$)	2·19	2·82	2·09	1·98
Porosity (%)	4·6	2·9	4·8	5·1
Unconfined compressive strength (MPa)	27·5	97·5	11·7	25·8
Point load strength (MPa)	2·1	3·7	0·3	0·6
Schleroscope hardness	27	38	12	9
Schmidt hardness	25	40	8	11
Young's modulus ($\times 10^3$ MPa)	24·8	63·9	3·8	7·9
Permeability ($\times 10^{-10}\,m\,s^{-1}$)	6·2	0·3	—	—

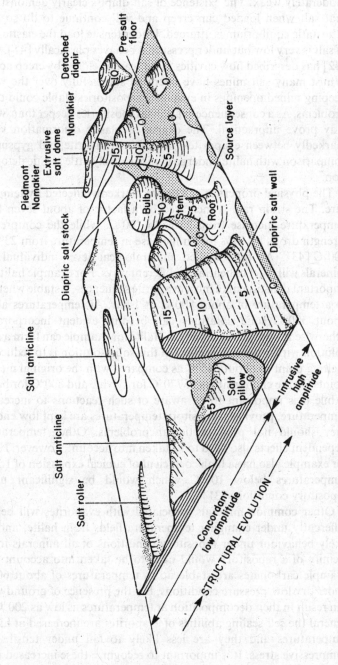

Figure 2.3 The main types of large salt structures. Structure contours are in arbitrary units. Salt nappes and irregular salt massifs have been omitted because they are an order of magnitude larger [35].

Labels on figure:

Detached diapir

Pre-salt floor

Namakier

Extrusive salt dome

Piedmont Namakier

Source layer

Diapiric salt stock

Bulb

Stem

Root

Diapiric salt wall

Salt anticline

Salt anticline

Salt pillow

Salt roller

Intrusive high amplitude

Concordant low amplitude

STRUCTURAL EVOLUTION

gypsum and potash are moderately strong, whilst rock salt is moderately weak. The existence of salt diapirs clearly demonstrates that salt when loaded can creep and will continue to do so until lithostatic equilibrium is attained. Under tensile load the elastic limit of salt is very low but under pressure it behaves plastically [41]. Baar [42] has described how cavities in salt can be closed by creep action. Whilst many salt mines have operated successfully over the years, keeping mined openings in evaporite repositories stable could cause problems. As a consequence evaporite repositories deeper than 800 m may prove unpractical. The degree of plastic deformation varies markedly between evaporites, however. Anhydrite and gypsum in comparison with halite undergo comparatively little plastic deformation.

The physical properties of salt are markedly affected by temperature. The strain rate is increased by a factor of about seven for a temperature increase from 23°C to 100°C, while the compressive strength drops by 10% for an increase in temperature from 23°C to 200°C [43]. At the detailed mineralogical level individual salt minerals will react to heating in different ways. For example halite, an important constituent of most evaporite sequences, is stable when dry to a temperature of about 800°C at 1 bar. At temperatures above about 300°C solid solutions can become evident incorporating otherwise isomorphous minerals. Halite for example can form a solid solution with sylvine. The effect of this solid solution is to reduce the melting point to about 645°C as compared with the original melting points for the end members of 770°C for sylvine and 800°C for halite. While it is important to be aware of such reactions to increasing temperature, provided repository temperatures are kept low enough they should not pose particular problems. Other temperature-dependent effects also need to be taken into account, however. Halite for example also has a high coefficient of cubical expansion of 1.21 at temperatures below 100°C which would be significant under repository conditions [31].

Other common minerals associated with evaporites will behave differently under changing temperature fields from halite, and the likely behaviour under repository conditions of all minerals in the vicinity of a repository would need to be taken into account. For example carbonates are stable up to temperatures of about 600°C under dry low-pressure conditions, but the presence of groundwater can result in their decomposition at temperatures as low as 200°C. In general the self-sealing abilities of evaporites are increased at higher temperatures and they are less likely to fail under tensile and compressive stress. It is important to recognize these increased rates of viscoplastic deformation accompanying the heating of salt as this may significantly affect the rate at which repository openings close up.

Problems associated with the heating of evaporite minerals need not necessarily arise, however, as when compared with other rock types evaporites display relatively high values for thermal conductivity compared with other rock types as shown in Table 2.4. A temperature difference, set up in the rock as the decay of radioactive material heats the rock around the waste repository to relatively higher temperatures than more distant rock, causes heat to flow through the rock. The rate of flow depends on the temperature gradient and the area and nature of the rock. Thus, in a state where all the temperatures are steady, the heat flow across area A is related to the thermal gradient by

$$\mathrm{d}Q/\mathrm{d}t = -kA\,\mathrm{d}\theta/\mathrm{d}s$$

where $\mathrm{d}Q/\mathrm{d}t$ is the rate of heat flow ($\mathrm{J\,s^{-1}}$), $\mathrm{d}\theta/\mathrm{d}s$ is the temperature gradient ($\mathrm{K\,m^{-1}}$) and k is a constant of proportionality which depends on the material. It is k which is referred to as the thermal conductivity. The minus sign indicates that heat flow is in the direction of decreasing temperature. The units of k are $\mathrm{J\,s^{-1}}$ $\mathrm{K^{-1}\,m^{-1}}$ or $\mathrm{W\,m^{-1}\,K^{-1}}$.

The thermal diffusivity of a rock may also be referred to. Thermal

Table 2.4 Thermal conductivities of rocks [25].

Material	Thermal conductivity ($\mathrm{W\,m^{-1}\,K^{-1}}$)
Earth's crust, average	1·67
Soil, dry	0·14
Clay (dry)	0·84–1·26
Clay (wet)	1·26–1·67
Shale	1·67–3·34
Gneiss	2·09–2·51
Granite	1·67–3·34
Basalt	2·18
Chalk	0·84
Limestone	2·09–3·34
Marble	2·97
Marl	2·09–2·93
Rock salt	3·34–6·28
Gypsum	1·3
Sand (dry)	0·33–0·38
Sand (10% moisture content)	1·26–2·51
Sandstone, dry	0·84–1·26
Sandstone, wet	2·09–2·93
Coal	0·13–0·3
Water	0·59

diffusivity a is related to the thermal conductivity k by the expression

$$a = k/C\rho$$

where C is the specific heat and ρ the density of the rock.

The high thermal conductivity of most salts would enable an evaporite deposit to transmit heat away from a HLW repository into the surrounding rock fairly rapidly, enabling repository temperatures to be kept within acceptable levels more easily than in many other rock types.

2.4 FLUID INCLUSIONS IN EVAPORITES

The presence of fluid inclusions and brines in halite and other evaporite minerals is also important. If these brines which are mostly composed of a mixture of potassium, sodium and magnesium salts are heated to between 250 and 280°C they can explode violently [43]. Even under normal mining conditions Chapman [31] refers to massive sudden rock bursts involving up to 7500 tons of salt.

The heating of fluid inclusions in halite may also cause them to migrate up the temperature gradient by a process of dissolution of salt on the hotter side, diffusion across the fluid and recrystallization on the cooler side. This phenomenon may have undesirable effects in a HLW repository. Firstly it may concentrate hot, potentially very corrosive brines around waste canisters; secondly, as the thermal conductivity of brine is lower than that of salt local hot spots' stress increases could develop around waste canisters; and, thirdly, it could cause fluids in the surrounding rock to move towards the waste thus increasing the chances of fluid–waste interaction. Brookins [10] however, believes the potential problem of fluid migration should not be overstated. He refers to a dyke emplaced in an evaporite sequence near to the WIPP site in New Mexico. The dyke which was injected at a temperature of about 800 to 900°C, well above the temperature currently envisaged in evaporite HLW repositories, does not appear to have influenced fluid inclusions in the adjacent evaporites outside a 2 m zone from the dyke. Brookins [10] argues that were a direct extrapolation possible from laboratory and model studies large amounts of fluid should have been generated near the dyke but this was not the case. Hence caution must be used when extending theoretical and laboratory work into the field.

2.5 HYDROGEOLOGY OF EVAPORITES

While evaporites contain only small amounts of water as a general rule, it is important to realise that they do contain some water. Jockwer [44] has made a detailed study of the water and gas content of the Asse Mine in the Federal Republic of Germany. He found that most water found in the mine salts was derived from hydrated minor minerals such as polyhalite and kieserite, and intergranular water adsorbed onto the crystal boundaries. The amount of water contained in fluid inclusions was found to be very low, as is typical of North German salt domes. Figure 2.4(a) produced by Jockwer [44] shows the distribution frequency of the total water content of some 202 rock salt samples taken from the Asse salt mine. From this it can be seen that the water content is not constant but depends on the amount of minor minerals present containing water of hydration. Figure 2.4(b) shows the vertical distribution of water within a 300 m deep core taken from the same area. The gas contents noted by Jockwer in these salts are: H_2S within the range 0–5 p.p.m., CO_2 0–200 p.p.m., gaseous hydrocarbons 0–60 p.p.m., and thermally generated HCl by decomposition of rare and trace minerals 0–150 p.p.m.

The question of whether most of the water trapped in evaporites is meteoric or non-meteoric has been approached on the basis of trace-element composition and isotopic distribution. Both approaches suggest most of the brines in evaporites are non-meteoric and are in fact formation waters. Such findings are important as they suggest such evaporites are isolated from circulating groundwaters which have surface connections. The situation regarding the origin of waters in evaporites can be complicated, however, by the dehydration of minerals at low temperatures. Carnallite, which can be present in large quantities in evaporites, is a case in point. As potential sources of water such minerals should be avoided in site selection. At the Asse mine in the Federal Republic of Germany, for example, the presence of a thick carnallite-bearing unit has posed problems with respect to potential waste storage and goes to emphasise the importance of recognizing and identifying the great mineralogical diversity which can occur in evaporite deposits.

It may be possible to avoid the effects of heating fluid inclusions by selecting the driest evaporites. Porosity values for evaporites range from about 1–10% and in general such pore space as does occur is isolated with little or no interconnection between voids. The consequence of this is that to all intents and purposes pure evaporites may be considered impermeable. In fact the hydraulic conductivity of most salt deposits is immeasurably low. Nevertheless it is important to recall that salt deposits may be interbedded with dolomite, calcareous shales, anhydrite, limestone and even sandstone, all of which may have quite high values for hydraulic conductivity. Indeed some of these interbeds or adjacent rocks in the case of salt domes

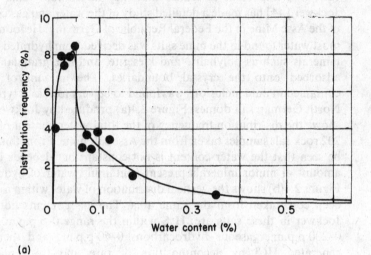

(a)

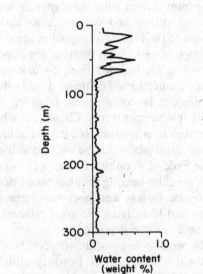

(b)

Figure 2.4 (a) Distribution frequency of the total water content of 202 rock salt samples from the Asse salt mine. (b) Water content profile within a halite sequence in West Germany. Note the very low water content and the way it varies from horizon to horizon. (Redrawn from [44].)

may even form aquifers. At the Gorleben salt dome site a water-bearing Quaternary channel formed originally across one branch of the salt dome has been identified [37], while in the Delaware Basin in south New Mexico, where the salt formations are quite near the surface, shallow circulating groundwaters have produced localized slumping [30].

Salt is quite soluble and evaporite deposits are potentially capable of dissolution on their margins where they come into contact with circulating groundwaters. Table 2.5 gives the solubilities of the more common evaporite minerals. The solution rate of evaporites is largely controlled by the surface area in contact with the water and the flow velocity associated with a unit area of the material. Evidence of dissolution at the top of salt domes is generally illustrated by the presence of a cap rock composed of a solid residue of less soluble materials left after the relatively much more soluble evaporite minerals have been dissolved out and transported away. Measurements of dissolution rates on the edges of the Mors salt dome located in the Jutland area of Denmark are given in Fig. 2.5. In the United States dissolution rates of between 5 and 15 vertical cm per thousand years have been estimated [45, 46]. Compared with the size of salt deposits under investigation and the likely depths of burial these relatively low rates of dissolution are unlikely to reduce significantly the viability of salt domes as potential HLW repositories.

If water gains entry to anhydrite beds subsequent hydration to gypsum can result in uplift. This comes about as a result of the 30–58% volume increase accompanying the hydration, which can generate pressures of as much as 69 MPa. The process of hydration may take place quite quickly and result in explosive release of

Table 2.5 Solubilities of evaporite minerals [31].

		30°C	100°C*	150°C	180°C	200°C
Halite	A	25·5	28·2	29·7	—	31·6
Sylvite	A	3·9	5·6	—	7·7	—
Gypsum	B	0·21	0·16	—	—	—
Anhydrite	A	0·56	0·15	0·02	—	0·008
Kieserite	A*	28	40·5	—	—	1·5

*Kieserite becomes hydrated to the hexa- or pentahydrate; both have similar solubilities.
A = units in g/100 g saturated solution.
B = units in g/100 ml saturated solution.
Polyhalite: $MgSO_4$ is the most soluble component. Data for a polyhalite + anhydrite mix at 55°C show a solubility of 18·92 g/100 g saturated solution.
Carnallite: is highly soluble, solubility being about 30 g/100 g solution at 35°C, little affected by increasing temperature up to 200°C.
Most of these data are for synthetic inorganic compounds.

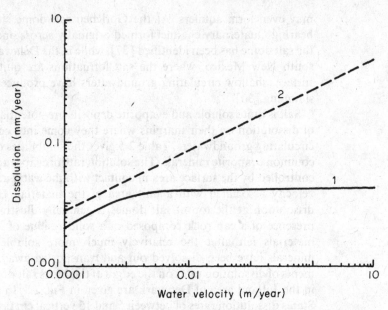

Figure 2.5   Calculated dissolution rates as a function of aquifer velocity. 1.
Stagnant water in a 20 m thick cap, porosity 10%. 2. Water
flowing between cap rock and salt mirror [38].

pressure where the anhydrite beds are buried at depth. Bell [40] notes
ground surface elevations of 6 m in association with the hydration of
anhydrite to gypsum. Water coming into contact with gypsum in only
small amounts and at low flow rates is unlikely to cause particular
problems. However, according to James and Lupton [47] if flow
rates exceed about 10^{-3} m s^{-1} extensive solution of gypsum is likely
to take place leading to the formation of caverns and enlarged joint
openings. Water gaining access to halite deposits can result in rapid
removal of salt material and the formation of cavities and surface
subsidence. Figure 2.6 shows the conditions for maximum hydrological stability and instability in salt domes.

2.6   THE RATE OF MOVEMENT OF SALT DIAPIRS

A particular problem with salt dome sites is the continued upward
migration of salt diapirs. As illustrated in Fig. 2.7, which shows the
genesis of the Mors salt dome, salt domes can migrate upwards
through considerable thicknesses of overlying strata [38]. In very dry
regions of the world such as Iran salt has been observed to flow out at
the surface. In southern Iran, Precambrian Hormuz salt has risen
diapirically from depths of 5–10 km through a predominantly

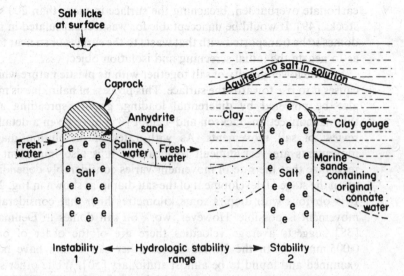

Figure 2.6 Contrasting conditions for maximum hydrological stability and instability. 1 is still showing vertical salt migration, 2 is proven to be tectonically stable, has no cap rock and no surface salt licks above it. (Redrawn from [48].)

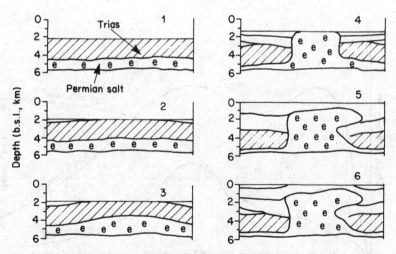

Figure 2.7 Development of the Mors salt structure. 1 200 million years before the present; 2 195 million years before the present; 3 190 million years before the present; 4 100 million years before the present; 5 60 million years before the present; 6 present. (Redrawn from [38].)

carbonate overburden, breaching the surface in more than 200 salt stocks [49]. It would be unacceptable for waste encapsulated in salt domes to be transported with that waste to the surface or so near to it as to negate the initial screening and isolation objectives.

It is the lower density of salt together with its plastic nature which causes it to rise towards the surface. This process of haliokinesis may also be powered by differential loading, gravity spreading and thermal convection. Jackson and Talbot [35] have given a detailed review of salt movements. As with rates of dissolution diapir movements are very difficult to measure. It is also important to recognize that the rate of movement varies considerably depending upon the stage of development of the salt diapir, as shown in Fig. 2.8. It is obvious with diapirs some kilometres high that considerable movement is possible. However, work on salt domes in Denmark [38] suggests average velocities there are of the order of only 0.005 mm y^{-1}. In the United States other salt domes have been examined and found to be almost stationary [50]. While other salt structures may be moving at much higher rates, with Vita-Finzi [51] for instance recording anticlines in the Persian Gulf rising at rates between 1.4 and 1.9 mm y^{-1}, and accepting that measurements of vertical migration mainly by reference to sediment thinning over salt domes involves assumptions and large uncertainties, the available work suggests there are salt diapirs stable enough to be compatible with the needs of a HLW repository.

Salt movement can also be expressed in terms of strain rate, which

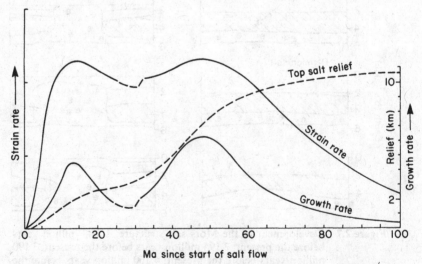

Figure 2.8 The life of a typical United States Gulf Coast salt structure. (Modified from [35].)

is the proportional change in length per second. Strain rates for salt vary widely reflecting the many different types of flow. Measured strain rates for *in situ* deformation of salt vary between 10^{-8} s^{-1} to 10^{-16} s^{-1}. The most rapid peak rates are for borehole closure at 10^{-8} s^{-1}. Mined cavities have been measured closing under strain rates of 10^{-9} to 9×10^{-12}. By way of comparison the average strain values for the crustal orogeny are about 10^{-14} s^{-1} [52].

2.7 THE EFFECTS OF RADIATION ON SALT

Bradshaw and McClain [43] have provided an account of the effects of radiation on salt. Radiation can result in a disordering of the crystal lattice in halite with a resultant storage of energy which can be released steadily as heat or explosively. Jenks and Bopp [53, 54] have shown that this stored energy is mainly dependent on dosage and the temperature of the salt at the time of irradiation. Below 120°C up to 60 cal g^{-1} can be stored. However, as the temperature rises less energy is stored until by 150°C only a negligible amount is stored in the salt. The conclusion here seems to be that if the salt temperature can be maintained above 150°C then energy storage problems associated with its irradiation can be avoided. A further consideration with irradiated salt is that it dissolves more rapidly; however, there is little reliable data available on this at present [31].

2.8 CONCLUSIONS WITH RESPECT TO EVAPORITES

Because of their plastic response to stress and tendency to close over openings it is unlikely that repositories much deeper than 800 m would be practical in evaporites. However, this self-sealing tendency, particularly at higher temperatures, and their ability to cope with tensile and compressive stress can be considered as positive assets in the context of utilizing evaporites as HLW repositories. It will be necessary, however, to have a clear idea of how various HLWs will affect closure rates and temperature profiles around potential repositories. While most evaporite minerals are quite stable at high temperatures when dry there are exceptions. Polyhalite and kieserite, for example, are unlikely to be suitable for high-temperature repositories as they both lose water of crystallization at temperatures between 300 and 450°C. Halite and anhydrite, on the other hand, owing to their anhydrous nature are not susceptible to change until much higher temperatures are reached. In general, evaporites to all intents and purposes are impermeable and the simple presence of these highly soluble minerals goes some way to suggest they are not generally associated with circulating groundwaters. Attention still needs to be given, however, to the possibility of the marginal dissolution of evaporites.

The presence of fluid inclusions in otherwise stable evaporites, notably halite, can cause problems above 250°C when rapid fluid decrepitation may occur. It may be wise to suggest an upper design temperature limit of between 150 and 200°C for evaporite repositories with caution leaning towards the lower end of the range. This would significantly reduce the likelihood of adverse effects arising from the irradiation of the salt. There is the suggestion that dome salts may be quite stable; nonetheless it remains important to be able to measure existing, and predict future movements in diapiric evaporites if these are to be considered for HLW disposal.

3 *The suitability of crystalline rocks as HLW repositories*

3.1 INTRODUCTION

Crystalline rocks have attracted interest as potential HLW repositories owing to their great strength, resistance to weathering, relative dryness, ability to retain radionuclides within the rock mass, and their high melting point. Whilst a range of crystalline rocks are under investigation, such as gabbro in Sweden and Canada, diabase in Japan and basalt in the United States, it is principally the granites which seem best suited to HLW disposal and this is reflected in the granite research programmes now under way in some nine countries (see Table 1.22). This research effort into granites is indicative not only of their potential as waste repositories but also of their widespread global distribution. While at the detailed level individual crystalline rock types will differ, the study of the granites can be used to illustrate features common to most crystalline rocks.

The exact origin of granites is surrounded by much controversy but it is probably true to say they result from the crystallization of molten material which may have its origins in or beneath the Earth's crust. For the most part granites occur as extensive masses covering many square kilometres and may be six or more kilometres thick, forming large batholiths and laccoliths.

3.2 MINERALOGY OF GRANITIC ROCKS

In general terms granite can be said to be a coarse-grained igneous rock consisting essentially of quartz, two types of feldspar with the potassium type being more common than the plagioclase type, and mica which normally occurs as biotite and muscovite. A number of accessory minerals such as apatite, zircon and magnetite may also be present. Granites which have been affected by hydrothermal or pneumatolitic action may contain in addition many other minerals such as tourmaline, topaz, kaolin and metals, notably tin, zinc and iron. Table 3.1 gives average values for the major oxides found in granites. The relationship between the quartz and the two feldspars which go to form the bulk of most granites is shown in Fig. 3.1. The existence of such phase diagrams is important as they enable the

Table 3.1 The major oxides (%) found in granites.

SiO_2	77·0	Al_2O_3	12
Fe_2O_3	0·8	FeO	0·9
CaO	0·8	Na_2O	3·2
K_2O	4·9	H_2O	0·3
TiO_2	0·1	MgO	0·08
Others	0·1–0·2		

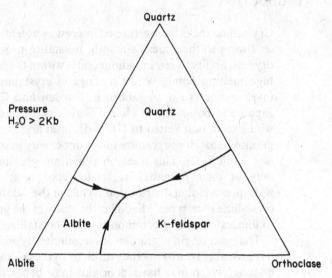

Figure 3.1 Schematic ternary phase diagram for granites.

geologist to predict likely mineral associations in granites and many other types of igneous rocks. However, as Table 3.2 shows the analyses for individual granites can vary widely along a more or less continuous series, while Fig. 3.2 shows the range of minerals likely to be encountered in other crystalline rocks beside granite.

Individual batholiths or other granite masses should not be thought of as being homogeneous, as not only is there wide variation in the granite itself but it may be cut by later igneous intrusions, or contain xenolithic blocks or fractures and joints filled with minerals derived from hydrothermal fluids or migrating groundwaters. Table 3.3 shows the great variety of minerals found in fractures in the Finnsjön granite in Sweden. Allard, Larson, Albinsson [57] state 'in granitic bedrock the mineralogy of water-carrying fractures where the predominant radionuclide transport would be expected is entirely different from the host rock itself.' A proper investigation of the detailed chemistry of cavity and fracture infills is therefore essential to determine whether the infilling material is derived from deep or

Table 3.2 Analyses of alkali granites [55].

	I Charnockite India	II Sodic leucogranite Mount Sorrel UK	III Riebeckite granite Nigeria	IV Aegirine granite Rockall	V Potassic leucogranite, Dartmoor UK	VI Potassic granite, Dartmoor UK	VII Potassi-sodic granite, Dartmoor UK	VIII Potassic granite, Lamorna UK
SiO_2	70·65	76·70	76·25	70·31	73·16	73·66	71·69	74·69
Al_2O_3	15·09	12·58	10·86	7·53	13·95	13·81	14·03	16·21
Fe_2O_3	0·80	0·10	1·23	8·32	0·03	0·21	0·57	tr.
FeO	1·53	2·09	0·76	2·44	0·47	1·51	1·93	1·16
MgO	0·53	0·65	0·18	0·02	tr.	0·45	0·66	0·48
CaO	2·66	1·10	0·37	0·35	0·43	0·67	1·49	0·28
Na_2O	2·99	4·90	4·68	5·26	2·57	2·89	3·03	1·18
K_2O	4·69	0·52	4·65	4·19	8·16	5·02	4·59	3·64
H_2O	0·65	0·85	0·50	0·43	0·64	1·66	1·76	1·23
TiO_2	0·46	0·20	0·11	0·26	0·04	0·16	0·33	—
Rest	0·10	0·14	0·36	0·54	0·17	0·31	0·29	0·68
	100·15	99·83	99·95	99·65	99·62	100·35	100·37	99·55

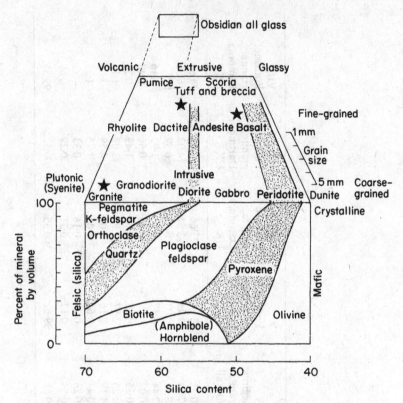

Figure 3.2 A classification of the igneous rocks. Rock types of particular interest as high-level nuclear waste repositories are starred. (Redrawn and modified from [56].)

surface sources and to assess its potential to adsorb radionuclides. If the infill material is derived from material shallower than the granite it may suggest potential links between the granite at depth and the biosphere with circulating groundwater acting as the intermediary. Brookins [10] illustrates this point in the context of deciding whether calcite infills in fractured granite rock at Fenton Hill, New Mexico were derived from overlying limestones or from some deeper source. By comparing the $^{87}Sr/^{86}Sr$ ratios for the overlying limestone and the calcite infills in the granite, Brookins concludes that the calcite is not derived from the overlying limestones, and that in this case the granite has been isolated from any contact with meteoric water.

3.3 PHYSICAL AND CHEMICAL PROPERTIES OF GRANITES

The minimum melting point of a granite at 1 bar is about 960°C, but

Table 3.3   Fracture minerals in granite [57].

Mineral	Mineral class	Occurrence*
Calcite	Carbonate	A
Dolomite	Carbonate	S, Me
Gypsum	Sulphate	P
Pyrite	Sulphide	A
Epidote	Soro silicate	Me, H, S
Prehnite	Soro silicate	Me, H
Chlorite	Phyllo silicate	A
Kaolinite	Phyllo silicate	W, H
Montmorillonite	Phyllo silicate	S, H
Illite	Phyllo silicate	W, H
Quartz	Tecto silicate	A
Laumontite	Tecto silicate	S, H
Stilbite	Tecto silicate	S,H
Analcime	Tecto silicate	S, H

*A = occurs in all geologic environments.
 S = in sedimentary rocks.
Me = in metamorphic rocks.
 W = as weathering products.
 H = as hydrothermal products.

at a depth of 3 km this will have fallen to about 250°C. The presence of fluids within the granite will, however, markedly alter its melting point. In consequence the greater the fluid content of a granite magma the greater chance it has of solidifying before it reaches the surface. This means that the majority of granites high in the erosional sequence are likely to have been dry magmas, evidence of which is seen in their lack of hydrous minerals such as mica. In contrast granites from deeply eroded areas are likely to contain a high proportion of hydrous phases [31]. The importance of this observation lies in the fact that at relatively high temperatures certain hydrous minerals in granites, especially pegmatite-rich blocks of rock, may be altered in such a way as to release their volatiles, a description of which is provided by Chapman [31]. While this is an aspect requiring consideration when deciding suitable repository design temperatures it is unlikely to be significant at temperatures below 250°C.

Granites – as their extensive use as building materials testifies – enjoy considerable strength, though this can be greatly reduced in weathered zones. However, contrary to what might intuitively be thought Blacic [58] has shown that in close proximity to a granite HLW repository creep deformation of this hard rock is likely to occur, posing problems for repository design – an example, perhaps, of the necessity of examining all aspects of a potential repository and to avoid uncorroborated intuitive assumptions. Table 3.4 gives the

Table 3.4 Some physical properties of igneous and metamorphic rocks [40].

	Relative density	Unconfined compressive strength MPa	Point load strength MPa	Shore scleoscope hardness	Schmidt hammer hardness	Young's modulus ($\times 10^3$ MPa)
Mount Sorrel Granite	2·68	176·4	11·3	77	54	60·6
Eskdale Granite	2·65	198·3	12·0	80	50	56·6
Dalbeattie Granite	2·67	147·8	10·3	74	69	41·1
Markfieldite	2·68	185·2	11·3	78	66	56·2
Granophyre (Cumbria)	2·65	204·7	14·0	85	52	84·3
Andesite (Somerset)	2·79	204·3	14·8	82	67	77·0
Basalt (Derbyshire)	2·91	321·0	16·9	86	61	93·6
Slate* (North Wales)	2·67	96·4	7·9	41	42	31·2
Slate† (North Wales)		72·3	4·2			
Schist* (Aberdeenshire)	2·66	82·7	7·2	47	31	35·5
Schist†		71·9	5·7			
Gneiss	2·66	162·0	12·7	68	49	46·0
Hornfels (Cumbria)	2·68	303·1	20·8	79	61	109·3

*Tested normal to cleavage or schistosity.
†Tested parallel to cleavage or schistosity.

main physical properties of the igneous rocks. To a certain extent, as Table 3.5 shows, certain of these physical properties may alter with changing temperature. The strength of granite and the experience of mining it to recover tin and uranium amongst other minerals should, however, make it possible to design and construct stable repository openings. Mines in granites could be readily constructed with existing technology to depths of 1000 m or possibly more [59].

Table 3.5 Temperature dependence of properties for granite [58].

Temperature (°C)	Thermal conductivity $(W\,m^{-1}\,°C^{-1})$	Specific heat $(J\,kg^{-1}\,°C^{-1})$	Linear coefficient of thermal expansion $(10^{-6}\,°C^{-1})$	Young's modulus (GPa)	Poisson's ratio
15	3·00	800	8·0	40	0·20
25	2·86	814	9·1	39	0·20
50	2·74	827	10·2	38	0·19
100	2·51	855	12·4	34	0·17
150	2·31	881	14·6	30	0·13
200	2·14	910	16·8	25	0·10

Notes: 1. Density of granite is assumed to be independent of temperature.
2. Properties of backfill are assumed to be temperature-independent.

Fractures and joints in granites are common though at depth these tend to be tightly closed. This is shown by water yield tests from wells in crystalline rocks which show well yields decreasing rapidly with depth [58, 60]. The decrease in hydraulic conductivity with depth is clearly demonstrated in Fig. 3.3, which relates to a Swedish gneiss/granite complex. Fractures and joints, however, are often concentrated in narrow zones and may have common preferred orientations with the result that large blocks of low-permeability granite may be bounded by zones of extensive fracturing with a markedly higher potential to allow the circulation of groundwater. Work on French granites by Barbreau, Bonnet, Goblet [60] using a 1000 m deep borehole also demonstrates that permeability in granites can be very low, in this case falling to about 10^{-12} m s^{-1} below 500 m. However, it is also worth noting that the highest values of permeability, about 10^{-9} m s^{-1}, occur near levels 300 m and 600 m which the authors associate with a small group of localized fractures.

Major fissures in crystalline rock may in turn be connected to an extensive network of microfissures as shown in Fig. 3.4. These microfissures are important as they provide a means by which radionuclides escaping from a repository diffuse into the rock matrix where they may be effectively immobilized over quite short distances.

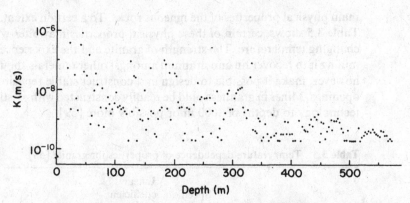

Figure 3.3 Hydraulic conductivity in drillhole Ka2, Sterno, Sweden. Data obtained by straddle-packer water injection tests. Distance between packers 3.0 m, K=hydraulic conductivity. Note how hydraulic conductivity decreases with depth [61].

Work by Neretnieks [63] and Grundfelt [64] suggests that diffusion of radionuclides into granite microfissures would result in the decay of the important chains including ^{241}Am, ^{237}Np, ^{229}Th and ^{226}Ra to insignificant levels well within a 350 m thick granite barrier. The only three significant nuclides which would not decay 'totally' within the same 350 m are likely to be the long-lived ^{135}Cs, ^{129}I and ^{238}U.

Fractures and joints in granites may result from the cooling and

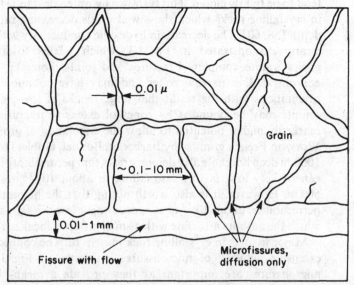

Figure 3.4 Two-dimensional view of microfissures in granite showing 'typical' sizes of grains, microfissures and fissures [62].

associated contraction they undergo following their emplacement in the Earth's crust, and from earth movements. However, as most of the granites under investigation for HLW disposal were formed between 250 million and 2000 million years ago they will have cooled to the appropriate temperature for their depth of burial, and further joint formation associated with cooling is unlikely. In this context it is of note perhaps that granite, like most igneous rocks, contains small amounts of radioactive minerals, generally averaged as about 4 p.p.m. of uranium, 13 p.p.m. of thorium and 4 p.p.m. of potassium, the decay of which produces heat of about $300 \text{ erg g}^{-1} \text{ y}^{-1}$. The comparable figure for basalt is about $50 \text{ erg g}^{-1} \text{ y}^{-1}$. In general pressure release and joint formation following removal of overburden during uplift of granites, like earth movements generally, is also unlikely if granites associated with the ancient shield areas are used. In these ancient continental heartlands rock movements are of the order of only a few mm per thousand years. It is reasonable then to assume that if the zones of known seismic activity are avoided in repository selection joints and fractures in the rock are unlikely to increase in number or size during the necessary lifetime of the site.

However, it is important to remind the reader that heat generated by the waste and subsequent cooling may alter the size and extent of existing fissures or form new ones locally. Figure 3.5 shows how for one set of circumstances the temperature and stress fields around a crystalline repository might be expected to change following the storage of HLW. It is also important that fracture zones are correctly detected and avoided in repository location. In this respect modern drilling techniques and experience gained in the drilling of deep granites in the respective hot dry rock programmes in the Fenton Hill area of New Mexico in the United States and at Camborne in the United Kingdom, together with the application of resistivity and magnetic susceptibility logging combine to suggest that fracture zones can with care be reliably detected [65]. Table 3.6, for example, shows the type of measurements possible in a 1000 m hole drilled into a French granite.

3.4 THE HYDROGEOLOGY OF GRANITES

Owing to their mode of origin it is unlikely that most granites would be intimately associated with sedimentary formations which can often form good aquifers. Certainly on their margins granites may abut country rock aquifers but in general the great bulk of granites would permit proposed HLW repositories to be located sufficiently distant from marginal aquifers. While underlying aquifers would not normally be expected in association with granites, though these could

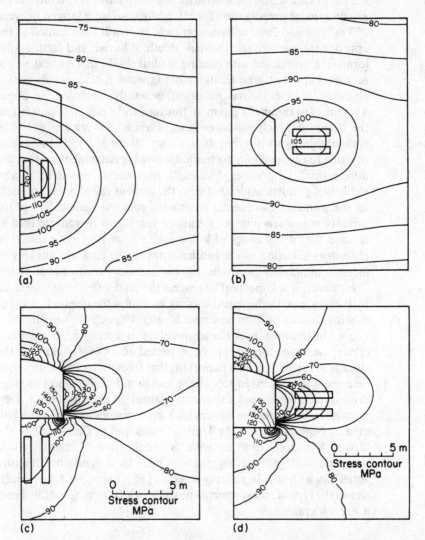

Figure 3.5   (a) and (b) show the estimated rise in temperature around a high-level nuclear waste room 30 years after emplacement of the waste. Temperature in °C with 5°C contours. (a) shows the situation for a vertical-borehole concept, (b) describes the horizontal-borehole concept. (c) and (d) show the combined horizontal stress around the repository after the same period. (c) relates to the vertical-borehole concept and (d) to the horizontal-borehole concept. (Redrawn from [58].)

Table 3.6 Deep borehole (1000 m) in granite measurements and results [60].

Investigated volume	Techniques	Concerned parameters	Some results
Cores in field	Orientation	Fracture frequencies directions, dipping	4–5 fissures/m (rather uniform density) Mostly vertical fractures but some horizontal fractures
	Structural analysis		
	Rock quality designation		
	Petrography	Nature of matrix	Homogeneous facies: 'monzogranite porphyroide'
Cores in laboratory	Spectral porosity by krypton and mercury method	ω = macroscopic porosity → specific surface	$\omega = 0\text{–}35\%$ $0\cdot10\text{–}0\cdot3$ m^2 g^{-1}
	Mineralogy	Alteration products → Kd of these products	Determined products: illite (50%), montmorillonite, quartz calcite
Open-hole logging	Electrical and nuclear diagraphy	Localization of fissured zones	Two main fractured intervals
	Smic diagraphy		$\dfrac{\text{Shear wave}}{\text{compressional}} = 0\cdot58$
Open borehole by isolated sections	Slug tests	Ke	$0\cdot50$ m $K = 6 \times 10^{-9}$ m s^{-1}; (slug) $60\text{–}250$ m; $K = 7 \times 10^{-11}$ m s^{-1}
	Injection tests: double packer	e variable from 10 m (double packer) to entire non-completed borehole	$250\text{–}540$ m; $K = 3\cdot4 \times 10^{-11}$ m s^{-1}; or $= 2\cdot7 \times 10^{-10}$ m s^{-1} (injection)
	simple packer		$540\text{–}1000$ m $K < 10^{-12}$ m s^{-1}

theoretically occur where granites take on a sheet-like mushroom form in their upper region, they may readily be overlain by aquiferous rock formations. In these circumstances it is important to establish what interconnection if any exists between such overlying aquifers and the granite groundwaters.

Groundwater within the granites themselves has been observed to be stratified with relatively younger and less saline waters overlying deeper older saline groundwater [66]. There is evidence from work carried out on Canadian granites [67] however that there can be interconnection between these shallow waters and other groundwater as deep as 400 m. At the Stripa mine in Sweden a slow downward and upward movement of groundwater has been identified [68] though the situation here is not entirely representative owing to alterations to the hydrologic regime by mining activity. It is also important to recognize that groundwater composition can vary markedly within any one granite complex. Allard, Larson, Albinson [57] have shown this variation in groundwater with analyses of samples from seven different boreholes down to depths of 500 m in the Finnsjön granite of Sweden as shown in Table 3.7. Differences in groundwater can have important consequences for waste canister corrosion and concrete mix for engineered barriers.

Table 3.7 Groundwater composition at Finnsjön [69].

	Concentration mg l^{-1}	
	a	*b*
Ca^{2+}	22–60	107–1790
Mg^{2+}	4–9·5	16–110
Na^+	13–124	224–1460
K^+	1·4–3·1	1·8–10·0
Fe(tot)	2·9–21	0·6–9·2
Fe(II)/Fe(tot)	0·8–1·0	0·25–1·0
Mn	0·05–0·31	0·06–0·74
Cl^-	13–124	380–5500
SO_4^{2-}	1–46	35–325
HCO_3^-	322–395	39–295
PO_4^{3-}	0·03–0·26	0·03–0·17
F^-	1·4–3·0	0·7–2·3
SiO_2(tot)	6·0–18	7·7–14
Organic C	6·2–11·0	1·2–6·2
Total	530–750	1300–9200
pH	7·1–8·8	7·7–8·4

a Fi1, Fi2, Fi4, Fi7:123 ⎫
b F7:301, Fi5   ⎬ Boreholes
                                        ⎭

Clearly if granites and crystalline rocks generally cannot be considered as dry it is necessary to have an understanding of how groundwater moves within such rocks; this is discussed in Chapter 8. It is also necessary to be able to predict how these groundwater movements will be altered by the emplacement of heat-generating wastes which will set up changing thermal and stress patterns. While it is probably true to say that fluid flow in porous media is now quite well understood the same cannot yet be said regarding our understanding of flow in fractured media. This is perhaps the principal area where further work is required before granites can confidently be considered as viable HLW repositories.

To a certain extent porous flow analogues can be used to predict groundwater flow in granites on the regional scale [70, 71], but on the local detailed scale porous flow concepts often break down. Recent work at the Stripa mine [72] and elsewhere tends to confirm that flow through large fractured networks can be reliably modelled. At the detailed level, however, work on individual fractures has shown that they cannot be considered as having smooth sides as previously assumed in many models and that flow is not uniform throughout fissures but is concentrated in isolated channels of enhanced flow. These recent advances in our understanding of fracture-dominated flow taken together with what was previously known has been succinctly reviewed by Gale [73]. Despite gaps in our knowledge it is considered far enough advanced for the NEA [7] to suggest that 'for many locations the current understanding of flow through crystalline rocks is adequate to predict with sufficient confidence the physical and hydrogeological behaviour of these rocks on both the small and large scale, and over the time intervals necessary in high level waste disposal.' Nevertheless, while in porous media it is often possible to measure the hydrogeological parameters of a formation at one location and then to extrapolate these reasonably confidently over a much larger area this is still inadvisable with fracture-dominated media and it is this perhaps which leads to the NEA [7] adding a note of caution to their previously quoted view as follows: 'however, it remains a difficult task requiring long-term monitoring to determine hydraulic properties at a particular site, and the level of confidence in the results will vary on a site-specific basis.' It is also important to recognize that shallow and deep groundwaters in granites may be in hydraulic continuity [67] and the heat generated by deposited waste may give rise to convective circulation of groundwater, moving deep water in the neighbourhood of a waste repository towards the surface contrary to the pre-waste-disposal groundwater flow regime, though in general it is likely that the natural downward regional flow of groundwater will be sufficient to overcome the buoyancy effects of thermal heating.

3.5   EFFECTS OF RADIATION ON CRYSTALLINE ROCK

Rankama [74] has shown that the physical properties of crystal structures can be altered by radiation of high-energy particles both by the physical alteration of the crystal lattice and by the production of free electrons and other changes. The physical properties of the minerals within a granite may be significantly altered by such radiation effects resulting in changes to density, hardness, unit-cell volume and brittleness. While heating can bring about annealing and repair of radiation damage because of the volume change which may be involved, minerals in the immediate vicinity of a repository may be shattered with consequent alterations to the hydrogeological properties of the rock. It is not intended to convey the impression that such effects are likely to be extensive (Jenks [75], for instance, has largely discounted irradiation effects in granites); nevertheless, the physical effects of radiation on the crystal structure of granites and other crystalline rocks need to be taken into account properly when considering potential HLW repositories.

3.6   BASALTS

These are much finer grained than granites and are dominated by mafic minerals and plagioclase feldspar. In the field they are normally seen to have formed in sheets or flows, though they can occur as vertical dykes or in other intrusive forms. The flows may range from a few metres to over 100 m in thickness, with individual flows having markedly different mineralogies and physical characteristics. For example, the Grande Ronde Basalt of the Basalt Waste Isolation Plant Site which straddles the Oregon–Washington state boundaries in the United States consists of 35 separate flows, ranging in thickness from about 7 m to 70 m and giving a total thickness of 800 m of basalt. The presence of rubble interflows and brecciation can lead to markedly differing values for porosity and hydraulic conductivity within any one pile of lava flows. Cooling may also induce the formation of polygonal, vertical or near-vertical fracture sets. Where sufficiently long periods elapse between eruptions sedimentary material may be laid down on old flow surfaces before being covered by new basalt material. While not particularly common it is therefore necessary to consider the possibility of finding sedimentary rock within piles of lava flows.

Basalts may be able to retard significantly the passage of radionuclides, through sorption on to contained zeolites, and montmorillonite clays derived from *in situ* chemical changes on the surfaces of rock fractures.

Basalts are hydrogeologically very complex, largely as a consequence of their varied physical characteristics and layered nature. Dense-flow interior basalts tend to have very low values for porosity and hydraulic conductivity. In marked contrast the presence of interflows and rubble horizons together with fractures and joints, commonly up to 1 cm in open width, render other sections highly porous and permeable. Basalt aquifers in the Columbia Plateau area of Washington, Oregon and Idaho have yielded flows of between 60 and 120 litres per second from wells up to 300 m deep. These yields are derived from a number of confined aquifers which coincide to a certain extent with individual basalt flows. However, vertical joints common in basalts allow water to move vertically through lava piles providing unconfined conditions in places. As Table 3.8 shows groundwater composition in basalts can also be very variable. Clearly groundwater considerations will play an important part in the assessment of a potential basalt repository.

3.7 TUFFACEOUS ROCKS

Tuff is primarily indurated volcaniclastic ejecta deposited either directly or reworked and redeposited by other surface processes. In its unconsolidated form such material would be known as ash. Tuffs have a very varied make-up but are generally recognized as falling into one of three groups: vitric tuffs, which are characterized by an abundance of glassy materials and shattered pumice; crystal tuffs, formed of mineral crystals, the mineralogy being determined by the nature of the original magma; and lithic tuffs which are marked by a dominance of volcanic rock fragments. The matrix material in most tuffs tends to be predominantly glassy ash. Welded tuffs may also be

Table 3.8 Groundwater composition in the Pasco Basin (modified from [76]).

Depth		> 750 m	300–500 m	0–300 m
pH		9·7–10·1	7·6–8·8	7·3–8·3
SiO_2		105–120	54–75	39–50
Cl		100	4–81	3–6
Na		180–240	30–150	5–21
HCO_3		2–51	144–277	85–186
CO_3	$mg\,l^{-1}$	100–127	?–15	?–
K		3·3–5·9	7·7–17	1·5–5·2
Mg		?–2	?–11	7–13
Ca		0·6–1·3	1–29	23–24
SO_4		10–96	?–27	10–18

encountered. These are a special group deposited at very high temperatures, generally above 500°C, which welded the various ash components together. Analyses of three tuffaceous rocks given by Pettijohn [77] are reproduced in Table 3.9. Post-depositional change in tuffs is seen in the formation of zeolites and various clay minerals which combine to give relatively high sorption capabilities to tuffs.

The physical properties of tuff vary considerably. Unwelded tuff for example can be less than one-third of the strength of welded tuff. The data given in Table 3.10 [79] give an order of magnitude for the tuffs under investigation as HLW repositories in the Nevada Test Site in the United States of America. Similarly the porosity and hydraulic conductivity of a tuff varies according to whether it is welded and upon the degree of welding. Porosity values for welded tuff are of the order of 10–30%. As a rule of thumb it has been suggested that the porosity of tuff is inversely proportional to the degree of welding, as measured by tensile strength. Tuffs also have a well-developed ability to retain large quantities of water within their available pore space even when situated above the water table. A representative ground-water analysis for the Nevada Test Site tuffs is given in Table 3.11.

Table 3.9 Representative chemical analyses of volcaniclastic sediments (modified from [77]).

Constituent	Pyroclastic deposits		
	A	B	C
SiO_2	70·40	53·63	48·67
TiO_2	0·21	0·96	1·99
Al_2O_3	13·65	19·59	14·15
Fe_2O_3	1·18	5·70	9·07
FeO	1·81	n.d.	0·83
MnO	0·04	—	—
MgO	0·07	3·35	6·36
CaO	1·58	3·53	6·16
Na_2O	3·76	3·64	1·61
K_2O	3·90	1·62	0·96
H_2O^+ } 4.03		7·91 }	9·39 }
H_2O^-		—	—
P_2O_5	0·06	—	0·36
CO_2	—	—	—
Total	100·69	99·93	99·5

A. Rhyolitic tuff, vitrophyre from John Day Formation (Middle Oligocene–Lower Miocene).
B. Andesitic tuff, Saleijer Island, Celebes.
C. Basaltic tuff, Szentgyorghegy, Zala, Hungary.

Table 3.10 Tuff: physical material properties (modified from [79]).

Property	Units	Intact rock
Young's modulus E	GPa	20
Poisson's ratio, v	—	0·25
Density, ρ	kg m^{-3}	2280
Thermal expansion coefficient	10^{-6} K^{-1}	7·5
		10·3
Angle of friction, peak	Degrees	43·2
Angle of friction, residual	Degrees	35·0
Cohesion, peak	MPa	8·5
Cohesion, residual	MPa	0·0

Table 3.11 Reference groundwater composition for tuff based on composition of Jackass Flats Well J-13 at the Nevada Test Site [80].

Constituent	Concentration mg/litre	Constituent	Concentration mg/litre
Silica	61	Lithium	0·05
Aluminium	0·03	Bicarbonate	120
Iron	0·04	Carbonate	—
Calcium	14	Sulphate	22
Magnesium	2·1	Chloride	7·5
Strontium	0·05	Fluoride	2·2
Barium	0·003	Nitrate	5·6
Sodium	51	Phosphate	0·12
Potassium	4·9		

pH 7·1.
E_h undetermined.

3.8   CONCLUSIONS WITH RESPECT TO GRANITIC AND OTHER CRYSTALLINE ROCKS

Granites, basalts and tuffs can generally be assumed to be strong enough to support physical excavation and the formation of suitable repository voids. However, it is necessary to recognize that granites vary considerably in their detailed mineralogy and that a predominance of hydrous minerals in a particular rock may render it liable to considerable change even at relatively low temperatures. In general, even in the absence of hydrous minerals it is likely that the design temperature of a repository in granite and other crystalline rocks should be kept below 200°C and, ideally, lower than this. The fracture-dominated nature of most crystalline rocks makes an appreciation of their hydrogeology very difficult to realize. This is particularly the case with respect to the layered nature of many basalts and tuffs, and over small distances can result in rapidly changing groundwater characteristics. The effects of thermal stress and other stress factors associated with the excavation of the rock itself can further complicate the nature of fracture-dominated flow paths in the immediate neighbourhood of a crystalline rock repository. These factors will need to be taken into account in repository design, as will the possibility of the establishment of thermal convection cells within the rock, energized by heat derived from the deposited waste. While fracture zones with high values for hydraulic conductivity can occur and should be avoided, there is evidence to suggest that elsewhere small fractures and their infills, together with the presence of clay minerals and zeolites, can play an important role in retarding the migration of radionuclides.

4 *The suitability of argillaceous rocks as HLW repositories*

4.1 INTRODUCTION

The argillaceous rocks are a broad and often complex group of detrital sedimentary deposits which may be broadly classified as clays, shales, mudstones, siltstones and marls. They are of interest as HLW waste repositories mainly because of their relatively low permeability, good sorptive characteristics, low solubility, and their ability to act in a plastic manner, which makes them self-sealing in certain circumstances. Against these advantages must be set their often low structural strength which poses particular problems for repository design and construction, their low thermal conductivity which may result in unacceptably high temperatures in the immediate vicinity of repositories, and the dewatering of hydrous minerals in response to thermal loading. At present major studies of argillaceous rocks are under way in Italy, Switzerland, the United Kingdom, the United States of America, Japan and Spain, but perhaps the most advanced work has been carried out in the Mol area of Belgium (see Table 1.22).

4.2 ORIGIN AND COMPOSITION OF ARGILLACEOUS ROCKS

The argillaceous rocks mainly form from the deposition of clay particles and other detrital material in seas and lakes. They may also be derived from the *in situ* weathering of other rocks, such as the alteration of feldspars in granites to china clay, and from terrestrially deposited material giving rise to rocks such as loess and adobe, or the weathering of volcanic rocks to form bentonites. Figure 4.1 shows the sources of mud and major processes of mud deposition.

In general the argillaceous rock types can be considered as being substantially composed of clay minerals, but other minerals as shown in Table 4.1 may also be present. While these other minerals may be important – for instance zeolites can play a significant part in reducing the mobility of radionuclides – the argillaceous rocks of interest as HLW repositories are for the most part formed of clay minerals. These can be classified into four main groups: the kandite

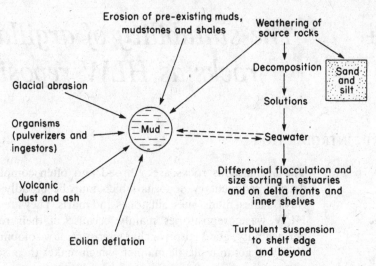

Figure 4.1 Sources of mud and major processes of mud deposition [81].

group, the illite group, the smectite group and vermiculite. Deer, Howie and Zussman [82] give the main characteristics of these clay minerals and describe them as having

'a number of characteristics in common. Chemically all are hydrous silicates which on heating lose adsorbed and constitutional water, and at high temperatures form refractory materials. Most of them occur as platy particles in fine grained aggregates, which when mixed with water yield materials which have varying degrees of plasticity.'

The structure of the principal clay groups is given in Fig. 4.2 from which it can be seen that the clays are made up of three different types of layered structures which combine in different ratios (shown in brackets at the base of each column). For the 1:1 clays such as kaolinite, for instance, the repeating unit is one layer composed of one SiO_4 sheet and one AlO_6 sheet. Important differences exist between the main clay groups, with for example, the water content ranging from as high as 22% by weight for the smectites to as low as 5% for the illite group. The kandites exhibit both the lowest cation exchange capacity and the highest anion exchange capacity of the main clay groups. The highest cation exchange capacity is to be found in vermiculite. When the main clay groups are combined in various proportions and mixed together with other detrital material the resultant rock types can be very complex, as the analyses given in Table 4.2 show. The presence of organic material in clay rocks is well documented owing to their believed significance in the production of hydrocarbons. These organic materials can have a significant impact

Table 4.1 Constituents of shales other than clays (modified from [81]).

Framework silicates

Quartz: Forms 20–30% of the average shale and is almost always present and is probably almost all detrital. May be, in part, aeolian in origin, but very little is known about its occurrence, largely because it is too fine-grained to study easily in thin section. However, the amount of quartz in a shale may be indicative of shoreline proximity. Other varieties of silica that may be present and should be looked for include chalcedony, opal CT, and amorphous silica, all of which may have originally been biological.

Feldspar: Nearly always less abundant than quartz, and plagioclase is believed to be more abundant than potash feldspar. Part of the feldspar in a shale may be authigenic, but little is known on how to distinguish very fine detrital from authigenic feldspar in shales. It appears that even less is known about feldspar in shales than about quartz.

Zeolites: Commonly present as an alteration product of volcanic glass, but can also be found in the muds of hypersaline lakes (analcite). Modern marine sediments have phillipsite and clinoptilolite as the most common zeolites, where they may form a small percentage of the mud. Zeolites are useful indicators of very low-grade metamorphism in shales (laumontite, 50–300 °C, and prehnite, as low as 90 °C).

Oxides and hydroxides

Iron oxides and hydroxides: Iron oxides or hydroxides are commonly present in shales, mostly as coatings on clay minerals. Such iron coatings are converted, in reducing environments, into pyrite or siderite. Hematite is the common iron oxide in shales, but in modern mud and weathered shales hydrous forms, such as goethite or limonite, are probably more common.

Gibbsite: The ultimate product of acid leaching, consists of $Al(OH)_3$. Characteristic of extreme tropical weathering, where it forms bauxites. May be associated with kaolinite in marine shale, the clay minerals of which have been derived from the weathering of a tropical land mass.

Carbonates

Calcite: Probably more common in marine than non-marine shale, but there should be no carbonate minerals of any kind in marine mud deposited below the calcite compensation depth (unless introduced by turbidity currents). However, as with quartz and feldspar, very little is known about the distribution and form of calcite in shale.

Dolomite: Appears to be common in shale, but its relation to calcite is unknown. Like calcite, it may be an important cementing agent.

Siderite and ankerite: Common in concretions. In marine shales, they indicate intermediate E_h values and so may, when compared with pyrite occurrences, be helpful in palaeogeographic reconstructions of those basins that have a more strongly reducing, deep part.

Sulphur minerals

Sulphates: Gypsum, $CaSO_4-2H_2O$, and anhydrite, $CaSO_4$ and barite, $BaSO_4$, occur as concretions in shale and may indicate some type of hypersalinity either during or after deposition.

Table 4.1 Constituents of shales other than clays (modified from [81]) (*contd*).

Sulphur minerals (*contd*)

Sulphides: The only abundant sulphides in shales are those of iron. Modern muds contain several amorphous varieties, but shales have only crystalline FeS_2, as pyrite or marcasite. Both are much more abundant in marine shales than in continental ones and both indicate strongly reducing conditions either at the sediment–water interface or within the sediment. Pyrite and marcasite are indistinguishable in the field and therefore their relative abundancies is not well known. There is some evidence that marcasite forms under lower pH conditions than pyrite.

Other constituents

Apatite: A phosphatic mineral that forms when surface waters have much organic productivity. Commonly apatite forms nodules in slowly deposited marine muds. Another phosphate mineral is vivianite, an iron-bearing phosphate, which may well be more abundant than we think.

Glass: Common in modern muds of either continental or marine origin that are associated with volcanism. Probably converted during burial to either zeolites or smectites and free silica.

Heavy minerals: Potentially legion and possibly best preserved in concretions, although little is known about their occurrence and abundance in shale.

Organic materials

Discrete and structured organic particles: These are mostly either palynomorphs or small coaly fragments (vitrinite). Very useful in Phanerozoic shales for correlation and can be used for helping to identify the thermal history of the basin.

Kerogen: Amorphous organic material that systematically changes colour with increasing temperature and finally converts to graphites. Present in almost all shales except red ones and tells about the gas and oil potential of a basin and its thermal history. Chemical characterization is complex.

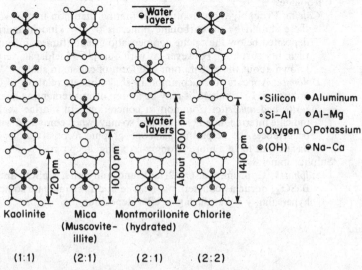

Figure 4.2 Schematic diagrams of some typical clays [2].

Table 4.2 Analyses of clay minerals [82].

	1	2	3	4
SiO_2	45·80	44·46	56·91	46·54
TiO_2	—	0·15	0·81	0·17
Al_2O_3	39·55	36·58	18·50	36·37
Fe_2O_3	0·57	0·36	4·99	0·72
FeO	0·18	0·07	0·26	0·36
MnO	—	—	—	0·00
MgO	0·14	0·18	2·07	0·50
CaO	0·41	0·19	1·59	0·22
Na_2O	—	0·01	0·43	0·46
K_2O	0·03	0·51	5·10	8·06
H_2O^+	13·92	13·38	5·98	6·31
H_2O^-	0·17	4·05	2·86	0·52
P_2O_5	—	0·18	—	0·06
Total	100·77	100·12	99·50	100·31
α	1·562	—	—	—
β	1·566	—	—	1·575
γ	1·568	—	—	1·580
D	—	—	—	2·65 ± 0·02

Numbers of ions
(on the basis of 18 (O, OH) for 1 and 2; 24 (O, OH) for 3 and 4)

	1	2	3	4
Si	3·94	4·01	7·50 } 8·00	6·00 } 8·00
Al	—	—	0·50	2·00
Al	4·01 ⎱	3·89	2·38	3·53
Ti	—	0·01	0·08	—
Fe^{+3}	0·04 ⎬ 4·08	0·02 ⎬ 3·95	0·50	0·07
Fe^{+2}	0·01	0·01	0·03 ⎬ 3·62	0·04 ⎬ 3·76
Mn	—	—	—	—
Mg	0·02 ⎰	0·02	0·41	0·10
Ca	0·04	0·02	0·22	0·02
Na	—	0·00	0·11 } 0·97	0·11 } 1·44
K	0·00	0·06	0·86	1·33
OH	7·98	8·04	5·26	5·45*

1. Kaolinite, hydrothermal veins of Cu–Pb–Zn ore, Niigata, Japan.
2. White halloysite, Bedford, Indiana.
3. Illite, Fithian, Illinois.
4. Hydromuscovite, Ogofau, Carmarthenshire, South Wales.
*Includes 0·02 F.

even at concentrations of 1% or less on the mobility of certain radionuclides. Studies in Black Sea sediments, for example, show a marked increase in the amount of uranium in seabed deposits as the amount of organic carbon in these sediments increases [83].

Special mention needs to be made of bentonite which owing to its relatively high cation exchange capacity has a potentially important role to play as a backfill and buffer medium within the multi-barrier concept. It is composed almost entirely of montmorillonite and colloidal silica, produced as the alteration product of glassy volcanic debris. As might be expected bentonite deposits frequently contain other clearly volcanic material such as idomorphic zircon or relict glass shards. Being the result of individual or closely related eruptions occurring over a brief period of time bentonite beds tend to be very thin and are normally less than 50 cm thick. As such they are not present in sufficiently thick sequences to act as repositories on their own.

In the field the following clay mineral rock types are recognized [84]: *Clay*, which is the term generally reserved for material which is plastic when wet and has no well-developed parting along the bedding planes, although it may display banding. *Shale*, which has a well marked bedding plane fissility, primarily owing to the orientation of the clay mineral particles parallel to the bedding planes. Shales do not form a plastic mass when wet, although they may disintegrate when immersed in water. *Mudstone* is a term used for rocks which are similar to shales in their non-plasticity, cohesion, and lower water content, but which lack the bedding-plane fissility. *Siltstones* are similar to mudstones but consist of a predominance of coarser silt-grade material, while *marl* is a calcareous mudstone. A classification of shale mudrocks is given by Potter, Maynard and Pryor [81] and this is reproduced as Table 4.3. A measure of the enormous variability that can occur in shale deposits is given in Table 4.4. It is important to recognize that, owing to the very variable nature of their mode of formation and the non-uniform combinations of minerals likely to be found in argillaceous rocks, unlike the igneous crystalline rocks, they cannot be treated as spatially homogeneous isochemical systems obeying basic petrological principles. Even on a small scale highly heterogeneous deposits can be encountered with the argillaceous rocks merging into certain of the other sedimentary rocks as shown in Fig. 4.3. It is also important to recognize that the principal shale minerals can undergo quite complex transformations as a result of chemical weathering, diagenesis and metamorphism, as shown in Fig. 4.4. Despite these factors, however, there have been identified a number of very thick almost isomorphous clay deposits which may reward further study. The Boom Clay (Tertiary) of the Mol area in Belgium, for instance, has a thickness of about 100 m and

Table 4.3 Classification of shale (more than 50% grains less than 0·062 mm) [81].

			0–32	33–65	66–100
Percentage clay-size constituents			0–32	33–65	66–100
Field adjective			Gritty	Loamy	Fat or slick
Nonindurated	Beds	Greater than 10 mm	Bedded silt	Bedded mud	Bedded claymud
	Laminae	Less than 10 mm	Laminated silt	Laminated mud	Laminated claymud
Indurated	Beds	Greater than 10 mm	Bedded siltstone	Mudstone	Claystone
	Laminae	Less than 10 mm	Laminated siltstone	Mudshale	Clayshale
Metamorphosed	Degree of metamorphism	LOW	Quartz argillite	Argillite	
			Quartz slate	Slate	
		High	Phyllite and/or mica schist		

Table 4.4 Chemical analyses of shales (excluding carbonaceous and siliceous shales) [77].

Constituent	A	B	C	D	E	F
SiO_2	25·05	51·38	49·85	56·73	58·82	58·10
TiO_2	—	1·22	1·45	0·88	0·73	0·65
Al_2O_3	8·28	23·89	13·88	19·27	·16·46	15·40
Fe_2O_3	0·27	2·05	3·75	5·57	1·10	4·02
FeO	2·41	5·01	14·10	1·89	7·20	2·45
MnO	4·11	0·02	0·24	—	0·09	—
MgO	2·61	2·71	3·32	1·93	4·92	2·44
CaO	27·87	0·24	0·20	0·01	0·76	3·11
Na_2O	—	0·59	0·10	0·49	4·03	1·30
K_2O	—	7·08	2·74	8·85	1·60	3·24
H_2O^+	2·86	4·66	4·90	3·77*	3·73 ⎱	5·00
H_2O^-	1·44	0·21	0·14	0·38	0·11 ⎰	
P_2O_5	0·08	0·01	0·09	—	0·17	0·17
CO_2	24·20	0·14	4·09	0·00	0·01	2·63
SO_3	—	none	—	—	0·02	0·64
S	—	none	1·51	—	0·05	—
C	—	0·16	0·69	—	—	0·80
Total	99·18	99·52	101·05	99·77	99·87†	99·95
Less O for S			−0·76		−0·06	
			100·29		99·81	

*'Loss on ignition'.
†Includes 0·03 Cl, 0·04 F.
A. Cretaceous shale, Mt Diablo, California. A very calcareous shale, also with abnormal MnO content.
B. Slate from the Tyler Formation (Precambrian), about 1 mile west of Montreal, Wisconsin. A high alumina rock, also potassic.
C. Dunn Creek slate (Precambrian), Homer Mine, Iron River, Michigan. C. Warshae. An iron-rich slate.
D. Average of six analyses of Cartersville Slate (Cambrian), Georgia. A very high potassic slate.
E. Varved argillite (Huronian), Olive Township, Ontario, Canada. Includes 0·03 Cl, 0·04 F. A soda-rich argillite.
F. Clarke's average shale for comparison.

lies between 190 m and 300 m below the surface, while the bentonitic mudstone Pierre Shale (Cretaceous) of the United States varies in thickness from 150 m to more than 1500 m and extends over 960 000 km^2.

Because they are sedimentary in origin it is usual to find argillaceous rocks interbedded with other sedimentary formations which may have markedly different physical and chemical properties

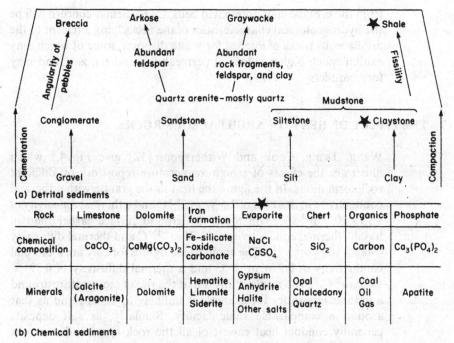

(a) Detrital sediments

(b) Chemical sediments

Rock	Limestone	Dolomite	Iron formation	Evaporite	Chert	Organics	Phosphate
Chemical composition	$CaCO_3$	$CaMg(CO_3)_2$	Fe-silicate -oxide carbonate	NaCl $CaSO_4$	SiO_2	Carbon	$Ca_3(PO_4)_2$
Minerals	Calcite (Aragonite)	Dolomite	Hematite Limonite Siderite	Gypsum Anhydrite Halite Other salts	Opal Chalcedony Quartz	Coal Oil Gas	Apatite

Figure 4.3 A classification of the sedimentary rocks. The starred rocks are those of most interest for the disposal of high-level nuclear waste. (Modified from [56].)

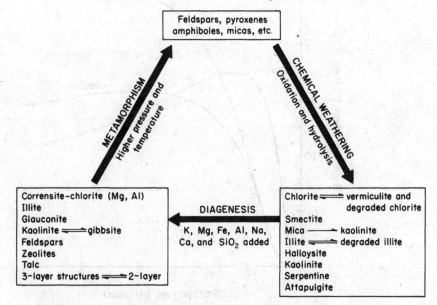

Figure 4.4 Mineralogic transformations and processes that generate the principal minerals of shales [85].

from those of the clay-dominated beds. Of particular concern will be the hydrogeological characteristics of the rocks lying adjacent to the argillaceous rocks of interest for waste disposal, some of which may exhibit much higher values of permeability and porosity and may form aquifers.

4.3   THE EFFECT OF HEAT ON ARGILLACEOUS ROCKS

Wang, Tsang, Cook and Witherspoon [12] give Fig. 4.5 which illustrates the effects of a high-temperature repository on different rock formations. In the figure the heat in the granite with a thermal conductivity in this case of 2.5 W m^{-1} °C and a thermal diffusivity of 1.15×10^{-6} m^2 s^{-1}, will reach the ground surface earlier than in basalt (thermal conductivity 1.62 W m^{-1} °C and thermal diffusivity 0.486×10^{-6} m^2 s^{-1}). The comparable figures for clay are a thermal conductivity of 0.9 W m^{-1} °C and a thermal diffusivity of 0.391×10^{-6} m^2 s^{-1}. The result of this is that the rise in temperature around a granite repository for example is unlikely to be as rapid as that around a comparable shale facility. Similarly, as salt deposits generally conduct heat easiest of all the rock types under serious

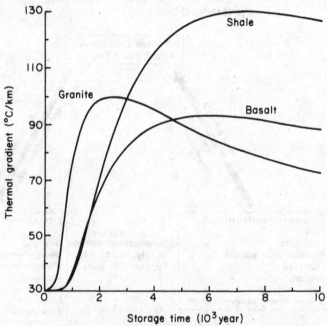

Figure 4.5   Thermal gradient rise at ground surface above centre of repository in different rock formations.

consideration, they would allow the most rapid transfer of heat away from a repository. This makes it easier to keep rock and groundwater temperatures as low as possible, which it is generally agreed is highly desirable. In clays, as with the granites and evaporites, it has been shown that heat dissipation occurs overwhelmingly by conduction, with the part played by moving groundwater being negligible [86].

Chapman [31] has highlighted how the difference in thermal properties of clays potentially poses difficult problems for the designer of a HLW repository. Younger clays which offer the desirable qualities of plasticity and low permeability often demonstrate poor thermal stability and pose difficult mining problems. On the other hand, whilst a compact older slaty unit has better thermal stability and improved engineering properties it is also likely to have a much higher fracture permeability and thus a greater degree of groundwater access. The clay mineralogy and hence the radionuclide potential of a shale may also be altered by diagenetic processes. With increasing depth of burial for example, kaolinite and montmorillonite generally decrease in abundance, with illite- and chlorite-related minerals taking their place.

Progressive heating of clays has been monitored at a number of sites, most notably in Belgium and Italy [87], and in the laboratory [88]. On heating to below 100°C about half of the structural and adsorbed water can be expected to be mobilized to combine with any fracture or pore groundwater which may be present. This fluid phase will contain significant quantities of dissolved salts and gases which may react with the rock itself or the waste containers within a repository. The NEA [7] are of the opinion that at the depths envisaged for disposal the likely temperatures would not be high enough to cause the vaporization of pore water. Experimental work suggests that this will not occur under a hydrostatic pressure of 2 MPa even at temperatures of up to 200°C [89]. At higher temperatures of up to 700°C, above which total melting may occur, the breakdown of the sulphates and carbonates and oxidation of the sulphides and organic matter is likely.

Where igneous material has been injected into clay rocks it is possible to assess the impact of high temperatures upon them. In Tuscany in Italy one such magmatic intrusion with an estimated initial temperature of about 800°C has been estimated as producing a temperature of 500–600°C in the clay. Mineralogical changes in the clay seem to be limited to within a 4 m zone of the intrusion while physical effects such as baking of the clay rock have been observed up to 12 m from the heat source [90, 91].

Aoyagi, Kobayashi and Kazama [92] have produced a facies classification for clay rocks based upon their clay content, age, depth of burial and by inference the temperature regimes they have been subjected to. The four main facies recognized by Aoyagi *et al.* are, in order of increasing grade,

1. montmorillonite + illite + chlorite with or without kaolinite,
2. hydrous illite + mixed montmorillonite with or without kaolinite,
3. illite + chlorite,
4. sericite + chlorite.

Type 4 is the highest grade and is the result of metamorphism. Type 2 exists up to about 80°C and 800 bars of pressure. Type 3 forms up to about 100°C and 2000 bars while above this further increase in temperature and pressure favours type 4 facies. While Chapman [31] quite rightly considers this to be an oversimplification of how clays will react to changing temperature regimes under repository conditions, it does to some extent provide a simple model for predicting the effects of a high-temperature repository on clay rocks.

4.4 THE HYDROGEOLOGY OF CLAYS

The porosity of freshly deposited clay sediments is very high, reaching as much as 50% or more. However, in undergoing compaction and diagenesis this level of porosity is markedly reduced. Hedberg [93] has shown that the porosity of a shale under an overburden of 1800 m or so would be only about 9%. In part it is possible to predict the porosity of shales as a function of their depth of burial as shown in Fig. 4.6. Tectonism and deformation can also contribute to further decreases in porosity. While porosity values of even 9% or so may still be considered high in the context of a HLW repository, however, owing to the very small size of the individual clay particles, generally less than 0·003 mm, water contained within the pores is generally rendered immobile by retentive forces. It is important to recognize the effect of the greater groundwater viscosity which may arise in clays. Under such conditions Darcy's law may not be valid as a means of predicting likely flow rates, especially where there are only low hydraulic gradients which may not be sufficient to produce water movement that might otherwise have been anticipated. Davis [95] has given a typical range of porosity values for consolidated and unconsolidated earth materials. From this it can be seen that rocks generally associated with readily recoverable quantities of groundwater such as sandstones and carbonates can have significantly lower porosity values than clays. This highlights the importance of hydraulic conductivity – in determining the rate of water movement through a rock and the volume stored. For a consideration of the flow of groundwater see Chapter 8.

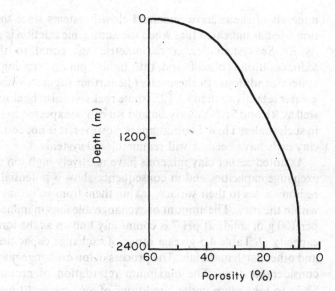

Figure 4.6 Relationship between porosity and depth of burial. (Redrawn from [94].)

The hydraulic conductivity of clays is generally very low, ranging between 10^{-7} and 10^{-11} cm s^{-1} as compared with values of 10^{-3} to 10^{-8} cm s^{-1} for sandstones. Well fractured igneous rocks may have values for hydraulic conductivity as high as 10 cm s^{-1} [96]. In general, then, unfractured clays can be considered as impermeable compared with most other rock types. It is important at this point, however, to repeat that argillaceous rocks can have very mixed natures and the presence of even small amounts of silt or sand-grade material can markedly increase the potential for groundwater movement. For example, the percentage specific yield of a typical silt deposit is of the order of 18% compared with 2% for a typical clay.

Hard, compact, brittle clays or shales may have much higher values for hydraulic conductivity than unmetamorphosed clays. Indeed, in Cornwall, England, there are many farms drawing water in considerable quantities from fractured Devonian clays. Heat generated by deposited waste will tend to drive groundwater away from the repository and this may move over considerable distances in well-fractured clays and may result in the creation of convection cells.

4.5 THE ABILITY OF CLAYS TO RETARD THE PASSAGE OF RADIONUCLIDES

Brookins [10] has addressed the question of whether the clay

minerals of shales have remained closed systems since their formation. He has indicated that when the authigenic fraction is studied for
its Rb–Sr systematics, a radiometric age equal to the age of
sedimentation is often found, thus in his opinion attesting to closed-
system conditions. In these cases he further suggests since Cs has a
greater retentivity than Rb [25] these rocks will also be closed to Cs as
well as Rb and Sr. Similarly Ba and Ra can be expected to be retained
in such shales. These findings apart, however, it is not certain that all
clay beds have been or will remain closed systems.

As noted earlier clay minerals have relatively high ion and cation
exchange capacities and in consequence show a potential to attract
radionuclides to their surface, taking them from solutions contained
within the rock. The amount of exchangeable ions in milliequivalents
per 100 g of solids at pH 7 is commonly known as the ion exchange
capacity Q. Table 4.5 gives a range of exchange capacities for shale
and other earth materials. The process of ion exchange may require a
considerable time. The maximum retardation of radionuclides is
likely to take place under conditions of long contact times between
solution and clay ion exchanger, slow water flow through the clay and
a high surface area of water–clay interaction. In general the reaction is
quickest in minerals in which the exchange takes place along the
crystal edges, such as kaolinite; but where the adsorption process also
requires the diffusion of the adsorbate into the adsorbent, as in
montmorillonite, for example, the exchange is much slower. In illite,
where the basal planes are firmly bonded together, the whole process
of ion exchange is even slower.

According to Matthess [25] cations are adsorbed onto clay
minerals in three main ways:

'1. On broken bonds around the edges of silica–alumina units,
preferentially on noncleavage surfaces (i.e. fractures). This type of bond
occurs particularly with kaolinite and hallyosite and with well-
crystallized chlorite, and sepiolite–attapulgite–palygorskite; for
smectites and vermiculites this makes up some 20% of the total cation
exchange capacity. The number of defect localities, hence the exchange
capacity, increases as the particle size decreases; for example in
kaolinite it increases fourfold from 10–20 μm grain size down to
0·1–0·05 μm; for illite an approximate two fold increase in exchange
capacity can be observed for grain size less than 0·06 μm compared with
grain size in the range 1·0–0·1 μm.
2. On equalization of unbalanced charges that originate by the
substitution of trivalent aluminium for quadrivalent silicon in the
tetrahedral sheets of some clay minerals, and trivalent aluminium by
ions of lower valence, particularly magnesium in the octahedral sheet.
The exchangeable cations in this case occur mostly on the basal
cleavage planes in the larger clay minerals. This type of bond is
responsible for about 80% of the exchange capacity of smectite and

Table 4.5 Exchange capacities of minerals and rocks (meq/100 g) from various data sets [25].

| Mineral Data sets | For cations | | For anions |
	'A'	'B'	'A'
Talc	—	0·2	—
Basalt	—	0·5–2·8	—
Pumice	—	1·2	—
Tuff	—	32·0–49·0	—
Quartz	—	0·6–5·3	—
Feldspar	—	1·0–2·0	—
Kaolinite	3–15	—	6·6–13·3
Kaolinite (colloidal)			20·2
Nontronite	—	—	12·0–20·0
Saponite	—	—	21·0
Beidellite	—	—	21·0
Pyrophyllite	—	4·0	—
Halloysite·$2H_2O$	5–10	—	—
Illite	10–40	10–40	—
Chlorite	10–40	10–40?	—
Shales	—	10–41·0	—
Glauconite	—	11–20	—
Sepiolite-attapulgite-palygorskite	3–15	20–30	—
Diatomite	—	25–54	—
Halloysite·$4H_2O$	40–50	—	—
Allophane	25–50	~70	—
Montmorillonite	80–150	70–100	23–31
Silica gel	—	80	—
Vermiculite	100–150	100–150	4
Zeolites	100–300	230–620	—
Organic substances in soil and recent sediments	150–500	—	—
Feldspathoids			
Leucite	—	460	—
Nosean	—	880	—
Sodalite	—	920	—
Cancrinite	—	1090	—

vermiculite and can have a certain importance in poorly crystalline illite, chlorite, and sepiolite–attapulgite–palygorskite. With the substitution bond the influence of grain size on the exchange capacity is hardly noticeable.

3. By replacement of the hydrogen in exposed hydroxyl groups, which are an integral part of the structure but are found along broken edges of crystals of clay minerals. This process is particularly important in kaolinite and hallyosite on account of the presence of an hydroxyl sheet on one side of the basal cleavage planes. Hydroxyl interlayers are important in montmorillonite and vermiculite.'

The rate of sorption will also vary according to E_h, pH, temperature and the mix of radionuclides, their relative solubilities and the presence of other material in solution. As Table 4.5 shows organic substances can also display relatively high ion-exchange capacities. In organic-rich clays which occur quite frequently this can be an important consideration, though at the same time the presence of organic material may give rise to acid groundwaters with increased corrosion potential of waste containers.

Figure 4.7 shows how pH can affect the sorption of different radionuclides, while Fig. 4.8 illustrates how the sorption of Sr is altered in the presence of Ca^{2+}, despite the fact that strontium is better adsorbed than calcium on clay minerals. Figure 4.9 shows clearly how different clay minerals may display markedly differing values of the distribution coefficient (K_d), which expresses the ratio between the concentration taken up by a unit weight of solid sorbent

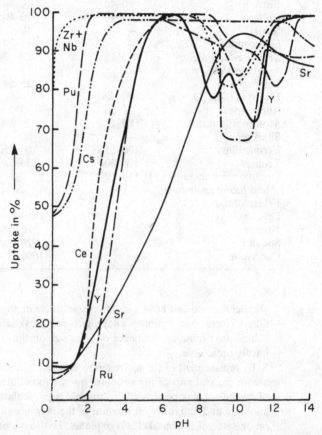

Figure 4.7   Influence of pH on the sorption of different radionuclides [25].

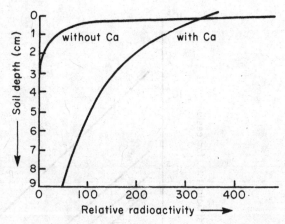

Figure 4.8 Influence of Ca^{2+} content in sandy humus-rich soil on ^{90}Sr sorption [25].

in chemical equilibrium with the concentration present in a unit volume of liquid. The distribution coefficient is a direct measure of the retardation factor any chemical species will undergo when conveyed by an aqueous solution or when moving in a continuous water matrix. An idea of the influence of temperature on sorption rates is given in Table 4.6 from which it can be seen that sorption generally reduces quite rapidly with increasing temperature.

The type of clay material involved will also be very important. The Boom clay of Belgium, for instance, contains vermiculite and smectite clay minerals which make it an excellent sorbent for various radionuclides with an average cation exchange capacity of 0.30 ± 0.05 meq g^{-1} [89]. In the Boom clay Sr may be expected to migrate up to 6 m, Cs 2 m, Eu 1–2 m, Pu up to 5 m. The long-lived fission

Table 4.6 Influence of temperature treatment on the distribution coefficient of clay (pH = 6 to 8) [89].

Nuclide initial concentration (0.1 mg l^{-1})	$T(°C)$	K_d (cm^3 g^{-1})
Sr	100	527
	300	208
	500	101
Cs	100	4900
	300	1229
	500	2414
Eu	100	3667
	300	2418
	500	229

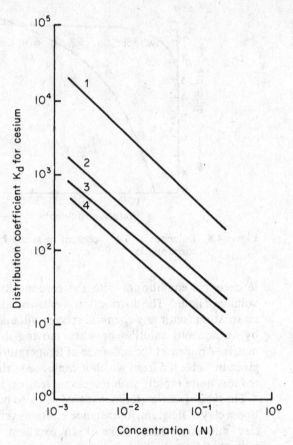

Figure 4.9 Distribution coefficient K_d for low concentrations of cesium in various clay minerals as a function of potassium concentration [25]. 1 potassium-montmorillonite; 2 potassium-illite; 3 potassium-kaolinite; 4 potassium-hallyosite.

products [129]I, [237]Np and ^{99}Tc are, however, practically not absorbed by clay minerals and provided sufficient time is allowed will diffuse out of even very thick formations. The rates of diffusion are, however, very slow and the level of radioactivity present on such diffusion flow paths reaching the surface from a deep repository is likely to be very low. The presence of organic remains within the clay in the form of carbonaceous material may locally increase sorption but at the same time the presence of organometallic constituents which are often associated with carbonaceous shales may increase the transport of radionuclides through these clays. As noted earlier while organic material can aid ion exchange the detailed role of organic material in clays and its effect upon radionuclide transport and sorption is an area requiring further study before conclusions can be reached on whether the presence of organic material is of benefit or not in clays to be used as HLW repositories.

4.6 CONCLUSIONS WITH RESPECT TO ARGILLACEOUS ROCKS

While the very plastic clays offer real advantages in terms of isolating buried waste from circulating groundwater they may not have sufficient strength to facilitate the economical construction of deep repositories. The self-sealing properties so attractive to the waste-repository designer pose real difficulties in keeping voids open at depth. The harder, stronger shales on the other hand are more structurally stable, but the presence of a fracture porosity in general means they have a higher hydraulic conductivity and allow for the easier passage of groundwater. Clays have relatively low values for thermal conductivity compared with the other rock types under consideration as HLW repositories, and this will limit initial design temperatures for repositories in clay, probably to temperatures of 100°C or less. On the other hand, clays have very high sorptive capacities. While it is difficult to be precise about just how much of a particular radionuclide would be retained by a clay repository should leaks occur, the existence of this property in the rock surrounding a repository is very desirable. It needs to be remembered, though, that sorption tends to decrease quite rapidly with rising temperature, a further reason to keep repository design temperatures at or below 100°C if possible. Equally important is the relative abundance of clays around the world, but considering the enormous variability which can occur both between and within clay strata, totally suitable units may be difficult to identify.

5

The containment of radionuclides within repositories

5.1 THE MAJOR PHYSICOCHEMICAL PROCESSES INVOLVED IN RADIONUCLIDE RETARDATION

The multi-barrier concept of containment referred to earlier relies on a systems approach to preventing radionuclides or radioactivity entering the biosphere following the disposal of HLW. We have seen how the characteristics of the three most likely geological host rock types can in part be amenable to the isolation of high-level nuclear waste, and various examples of how even when radionuclides escape from the immediate vicinity of a repository a number of physico-chemical processes can retard further radionuclide migration. A summary of those processes which play an important role in the systems approach to the containment of HLW waste is given in Table 5.1.

The NEA [15] and Skytte-Jensen [97] have summarized our present knowledge with respect to these processes. The retardation of radionuclide migration will depend to a large extent on the nature and specific site characteristics of the host rock. Not all the major physicochemical processes will be applicable to every type of earth material, though the first three listed in Table 5.1 can be expected to occur in all host media. No single process is always dominant, the

Table 5.1 The major processes involved in radionuclide retardation [7].

Chemical dissolution or precipitation of a solid phase
Diffusion into or out of adjacent flow paths under concentration gradients
Diffusion into the solid matrix of the host rock
Diffusion into fluid not involved in the bulk flow of the groundwater
Ion exchange with naturally occurring species
Direct sorption on solid surfaces
Chemical substitution reactions in which radionuclides replace stable isotopes of a different element in a stable phase
Isotopic exchange with the same element in a solid phase
Ultrafiltration of large-effective-diameter radioactive solutes or radioelement-bearing colloids by the host medium

---

reality being extremely complicated with various nuclides and processes interacting with each other and the host medium. It is important to recognize this complexity, for while it is possible to model many simple systems reliably (see Ames and Rai [98] and Muller, Langmuir and Duda [99] for example) the transport of reactive multi-component solutions of radionuclides in porous media cannot be fully modelled at present because of data limitations [7].

5.2 THE STUDY OF NATURAL ANALOGUES

While there are clearly advantages in studying physicochemical processes in the laboratory, it would generally be fair to say that the results of these experiments and tests cannot normally be reliably extrapolated far enough into the future to cover the likely periods envisaged for waste isolation. This shortcoming in our understanding may in part be rectified by a study of natural analogues. These are rock sequences containing naturally occurring radionuclides, the migration and stability of which can be measured after the passing of millions of years. Perhaps the best known of these analogues is the Oklo natural reactor which underwent a fission reaction for half a million years some two billion years ago. Brookins [100–103] has closely examined the Oklo rocks, which are essentially uranium-rich sediments. He has made careful note of which elements have migrated and the distances travelled where possible. Brookins [10] has also been able to attempt predictions of radionuclide migration at Oklo on the basis of E_h–pH diagrams. Table 5.2 summarizes the elements studied at Oklo and compares the predicted results with the actual migration patterns observed in the rocks. It would be wrong despite the considerable amount of work already carried out at Oklo to place undue emphasis on the results of that study, as much remains to be done. Nevertheless the available data suggest that rock can severely restrain the movement of radionuclides and acutely limit the distances covered where movement does take place, and that these movements can be predicted where good field data are available.

In support of the Oklo conclusions are the results of analogue studies of other rock types, mainly carried out in the United States of America. Laul and Papike [104] have studied chemical migration between granite and silt carbonate systems. Williams [105] has examined tuff and andesite systems, while Brookins [106] has studied the emplacement of a lamprophyre dyke into evaporites. All these studies apparently underline the findings of the Oklo studies showing only very limited migration of radionuclides if at all.

The importance of these analogue studies lies in their realism and tendency to represent conditions far more extreme than would

Table 5.2 Important elements at Oklo [10].

Element	Comments	E_h–pH predictions	
		25 °C	200 °C
Krypton	Most migrated	Not applicable	Not applicable
Rubidium	Probable local redistribution	Not applicable	Not applicable
Strontium	Probable local redistribution	Not applicable	Not applicable
Yttrium	Most retained	Retention	Retention
Zirconium	Most retained; some local redistribution	Retention	Some migration
Niobium	Most retained	Retention	Retention
Molybdenum	Most migrated	Migration	Migration
Technetium	Local redistribution	Retention	Migration (?)
Ruthenium	Local redistribution	Retention	Minor migration
Rhodium	Most retained	Retention	Retention
Palladium	Most retained	Retention	Retention
Silver	Most retained	Retention	Retention
Cadmium	Most migrated	Migration	Migration

Element			
Indium	Most retained	Retention	Retention
Tin	Not yet studied	Retention	Retention
Antimony	Not yet studied	Possible migration	Some migration
Tellurium	Not yet studied	Retention	Retention
Iodine	Most migrated	Not applicable	Not applicable
Xenon	Most migrated	Not applicable	Not applicable
Cesium	Most migrated (?; perhaps locally)	Not applicable	Not applicable
Barium	Local redistribution	Not applicable	Not applicable
REE	Most retained	Retention	Retention
Lead	Variable migration	Retention or local redistribution	Some migration
Bismuth	Most retained	Retention or local redistribution	Some migration
Polonium	Most retained	Retention	Retention
Thorium	Most retained	Retention	Retention
Uranium	Some local redistribution	Retention or local redistribution	Retention
Neptunium	Most retained	Retention	Retention
Plutonium	Most retained	Retention	Retention
Americium	Not measureable	Retention	Retention

probably be experienced in a man-made repository. Initial temperatures in these analogues are generally much higher than those
envisaged for HLW repositories, the time periods of isolation
observed are invariably far in excess of those required for waste
isolation, and they make no allowance for waste vitrification,
containment or buffering. While it remains important not to
transpose the findings of analogue studies recklessly to individual
repository sites and to recognise that high-temperature hydrothermal
fluids can penetrate through rock for considerable distances, natural
analogues do provide evidence to justify further examination of the
concept of geological disposal.

5.3 STABILIZING WASTE FORMS

At present most high-level waste is stored in liquid cooling tanks of
various design. While these are generally considered to be safe by the
nuclear industry they do have limitations and are not thought of as
providing a long-term solution to the disposal of high-level nuclear
waste [8]. As Table 5.3 shows the amount of radioactivity remaining
in stored waste remains high even after five years of storage while the
amounts of heat given off are considerable [107]. Thus, for example,
a 1000 tonne uranium fuel storage facility will have a radioactivity
inventory of the order of 10^9 Ci giving off about 12 MW of heat
during one year's cooling. Clearly, while pond storage is not regarded
as a long-term answer to HLW disposal it can have a marked effect
upon the radioactivity and heat-generating potential of the final
waste form. Figure 5.1 shows for example how various pre-geological
disposal storage times can greatly reduce the initial and maximum
temperatures reached, in this case in a granite repository.

   While it is possible to store liquid high-level wastes resulting from
the reprocessing of spent nuclear fuel the long-term disposal option

Table 5.3 Reduction in activity and heat release during storage in ponds of
LWR fuel (burn-up 37 000 MW d t^{-1}) [107].

Time after discharge from reactor (year)	Activity in fuel (Ci t^{-1} of U)		Heat release from fuel (kW t^{-1} of U)
	Total	^{131}I	
0·5	$3·9 \times 10^6$	0·17	
1	$2·7 \times 10^6$	$1·7 \times 10^{-8}$	12
2	$1·4 \times 10^6$	0·0	
5	$5·4 \times 10^5$	0·0	1·9

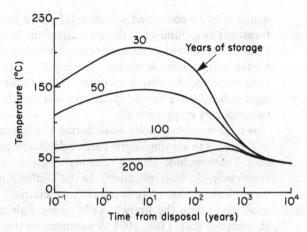

Figure 5.1 Maximum rock temperatures as a function of time for various pre-disposal storage periods [108].

most actively under examination remains deep geological burial. The purpose behind such burial is to isolate the waste from the biosphere for the requisite periods of time. As already noted the two ways in which waste might find its way back to the biosphere are by mechanical interference and by transportation in groundwater. It is generally agreed that movement by groundwater presents a more likely weak link than physical excavation. To dispose of waste as a liquid rather than a solid would undoubtedly make it more susceptible to groundwater transportation especially once the initial containment is breached. Table 5.4 gives the main technical requirements that a solidified waste form should satisfy.

Mendel, Nelson, Turcotte [110] have given a general review of these characteristics and their importance. Most importantly of these, thermal conductivity determines the heat production that can be allowed and therefore sets limits to activity concentration and dimensions of waste blocks, while leachability determines the rate at

Table 5.4 The main technical requirements of solidified high-level waste.

Low leachability
High chemical stability
High thermal stability
Resistance to radiation
Non-combustible
Low dispersal potential, i.e. powdered forms likely to be unacceptable
High melting point relative to the temperatures likely to be encountered in a repository, but not so high as to pose production problems
Mechanical stability

which activity contained within solid waste becomes available for transport by groundwater should it come into contact with the waste. Caution is required when considering and comparing leach-rate data for the various solid waste forms as leach tests are carried out under both static and dynamic conditions. In the former the leachate is not replenished, while in the latter the leachate is replenished either continuously or periodically.

A large number of solid waste forms which meet the criteria noted in Table 5.4 to varying degrees are under development and investigation. These include glasses, cements, calcines and various forms of synthetic rock, amongst others. In the United Kingdom and France borosilicate glasses have been taken to relatively advanced stages of production with the French AVM process already operating on a commercial scale [108, 109]. A summary of the characteristics of a number of solidified high-level wastes is given in Table 5.5. While the glasses have suggested the best option to date, attention has also focused on the use of an artificial rock known as SYNROC [111, 112]. SYNROC is a fine-grained rock similar in some respects to basalt. It contains minerals which when heated to 1200–1300°C are capable of fixing various radionuclides within the rock matrix. The actinides, for example, are incorporated into the zirconolite phase while cesium and barium are included in hollandite, and strontium in perovskite.

The rate of leaching of solidified wastes will vary mainly with time, temperature, groundwater flow rates and water chemistry. Figure 5.2, for example, shows leach rates for glass 209 varying with both temperature and water quality [108]. As temperature increases so the amount of material leached also increases. In consequence it is desirable to keep disposal temperatures below about 150°C. The effects of pH on leach rates can also be considerable with lower pH levels leading to higher leach rates as shown in Fig. 5.3 [10]. It should also be noted that high pH levels can similarly result in increased rates of leaching. Minimum leach rates take place in the pH range 6–8 and it follows that groundwaters in this range would be preferable to those showing more extreme values. A comparison of the leach rates of two glasses and a number of naturally occurring materials is given in Table 5.6. From this it can be seen that laboratory-derived leach rates for glasses compare favourably with many earth materials, though it is not necessarily the case that they share similar long-term stability.

5.4 THE WASTE CANISTER

After the waste form the next barrier to the transport of waste by groundwater is the waste container itself. It is generally envisaged

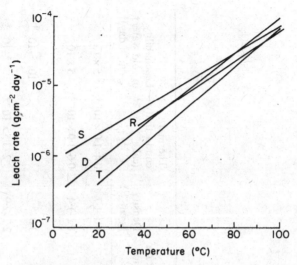

Figure 5.2 Leach rates for glass 209 under various conditions. S: sea water; T: tap water; D: distilled water; R: Severn river water, United Kingdom. Note how leach rate appears to be more sensitive to rising temperature than to original water quality. (Redrawn from [108].)

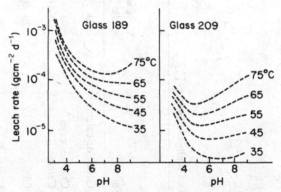

Figure 5.3 Effect of pH on leach rate of glasses at different temperatures. Note leach rate increases at extrene pH values [10].

that such containers will be cylindrical in shape, up to 4 m long and between 0·5 m and 1 m in diameter [113]. Thus, for instance, if earlier optimistic predictions of the United Kingdom's cumulative nuclear generated electricity of about 330 GW(e) by the year 2000 were realized no more than 2000 cylinders 3 m long and 0·5 m in diameter would be required to cope with the entire high-level waste resulting. Many materials have been investigated for use in the construction of

Table 5.5 Characteristics of solidified high-level wastes [1].

Solidification process	Source	Form	Wt% waste oxides*	Temperature of formation (°C)	Density (g cm⁻³)	Thermal conductivity (cal s⁻¹ cm⁻¹ °C⁻¹)	Leachability in cold water† (g cm⁻² day)†
Fluid bed calcination	USA	Granular calcine	50	400–500	$1{\cdot}0$–$1{\cdot}7$	$0{\cdot}4$–$1{\cdot}0\times10^{-3}$	5×10^{-1}
Pot calcination	USA	Calcine, cake	90	850–900	$1{\cdot}2$–$1{\cdot}4$	$0{\cdot}6$–$1{\cdot}0\times10^{-3}$	5×10^{-3}
Spray solidification	USA	Calcine	80–100	800			
Spray solidification	USA	Glass	20–40	1200	$3{\cdot}0$		10^{-2}–10^{-4}
Phosphate glass	USA	Glass	30	1200	$2{\cdot}7$–$2{\cdot}9$	2–$3{\cdot}5\times10^{-3}$	10^{-4}–10^{-5}
Rising level glass	USA	Borosilicate glass	30–50	900	$2{\cdot}9$–$3{\cdot}1$		10^{-5}–10^{-5}
FINGAL	UK	Borosilicate glass	25–40	1050	$2{\cdot}8$	$2{\cdot}5$–4×10^{-3}	10^{-5}–10^{-7}
HARVEST	UK	Borosilicate glass	25	900–1050	$2{\cdot}6$	$2{\cdot}75$–4×10^{-3}	10^{-5}–10^{-7}
PIVER	France	Borosilicate glass	20–30	1150	$2{\cdot}5$–$2{\cdot}9$	$2{\cdot}8$–$3{\cdot}6\times10^{-3}$	10^{-5}–10^{-7}
Continuous process (AVM)	France	Borosilicate glass	20–30	1100	$2{\cdot}5$–$2{\cdot}9$	$2{\cdot}8$–$3{\cdot}6\times10^{-3}$	10^{-5}–10^{-7}
VERA	FR of Germany	Borosilicate glass	20–30	1100–1200	$2{\cdot}5$–$2{\cdot}7$		10^{-5}–10^{-7}

Process	Country	Material	Waste loading	Temperature	Density	Thermal conductivity	Leachability
SYNROC	Australia and USA	Titanate ceramic	20	1150–1200	4·35	$3\ W\ m^{-1}\ k^{-1}$	10^{-1}–10^{-3} at 90°C
Pot solidification	India	Borosilicate glass	22–28	900–1050	2·5–3·0	$2·5$–$3·5 \times 10^{-3}$	10^{-6}
ESTER	Italy	Borosilicate or phosphate glass	20–25 / 20–25	1000 / 900	2·7–3·0 / 2·3–3·5		10^{-6}–10^{-7} / 10^{-5}
PHOTHO	FR of Germany	Phosphate glass	25–35	1000	1·8–2·9	$2·4 \times 10^{-3}$	$10^{-5} \times 10^{-7}$
LOTES	Eurochemic	Phosphate compound	30	450	2·1	$3·4 \times 10^{-3}$	10^{-6}–10^{-7}
PAMELA	Eurochemic /FR of Germany	Phosphate glass in metal matrix	25–35	Glass 1000°C incorporation in metal matrix 400°C		$2·5 \times 10^{-2}$	10^{-6}–10^{-7}
Nepheline syenite	Canada	Glass	0·1–3	1350			10^{-8}†
STOPPER	USA	Silicate or alumino silicate		Up to 500			10^{-11}§
THERMALT	USA	Aluminosilicate	25	2000			10^{-7}–10^{-8}
THERMITE	FR of Germany	Ceramic		2000	2·9		10^{-7}–10^{-8}

*Includes fission products and all other cations present in the waste solution, expressed as oxides.
†Leachability data are based on specific test methods and are time dependent. The indicated values are therefore not directly comparable.
‡Initially.
§Fifteen years later.

Table 5.6   Leach rates of rocks and glasses at 100°C [108].

Sample	Leach rate $g\ cm^{-2}\ day^{-1}$
Cotswold limestone	3.4×10^{-3}
Jurassic limestone	2.2×10^{-3}
Glass 189	1.3×10^{-3}
Basalt (Arthur's Seat Edinburgh)	4.0×10^{-4}
Glass 209	2.6×10^{-4}
Eocene Granite (Rockall Island)	2.0×10^{-4}
Pitchblende	1.0×10^{-4}
Granite (Scotland)	5.0×10^{-5}
Monzaite (Sri Lanka)	3.0×10^{-5}
Gneiss Diorite (Malvern)	2.5×10^{-5}

waste containers including stainless steels, cast irons, zirconium- and nickel-based alloys and titanium- and copper-based alloys [114].

It is not yet certain that containers can be developed capable of retaining their integrity for the very long periods of isolation required. A multi-layered metal container has been designed by the Brookhaven National Laboratory which should provide assured containment for a thousand years [115] but designs capable of containment for a million years or so are also under investigation [113]. According to Fyfe and Haq [116] advantage may be taken of the fact that some metals, such as copper, silver, gold and nickel, survive in rocks as small grains for billions of years. They take the view that two types of containers are theoretically capable of surviving for millions of years in low-permeability environments. A ductile copper container might be suitable for hard-rock sites, while aluminium oxide might be appropriate for salt deposits. In line with this Papp [19] reports work carried out in Sweden which suggests that copper containers capable of lasting intact for 100 000 years or longer can be constructed. Nevertheless, it is still felt to be prudent to assume that canisters will fail long before the need for containment is over. This apart, it is still necessary to develop canisters which will retain their integrity for at least as long as a repository is open. Advantages are also seen in the use of containers giving 1000 years or more of isolation as this would allow temperature and radiation effects to reduce substantially to a regime more easily simulated in the laboratory, so giving a greater degree of predictive control [117].

Lester [118] has considered how various waste packages might perform in practice; the results are given in Tables 5.7 and 5.8. The 'A' canister which is crushable is protected from cracking under the prevailing hydrostatic pressure by a 5 cm heavy iron overpack. The 'B' container relies for its strength upon the rigidity of the waste form

Table 5.7 Performance of waste packages in salt repository. Repository temperature=165°C; hydrostatic pressure=850 p.s.i.; lithostatic pressure=2500 p.s.i.; anoxic conditions [118].

Package type	Waste form	Heat loading (watts)	Crushable or rigid	Barrier element	Material	Thickness (in)	Time to waste-water contact (yr)	^{238}U release				^{99}Tc release			
								Breakthrough time (yr)	Steady-state release Begin (yr)	End (yr)	Rate (Ci yr^{-1})	Breakthrough time (yr)	Steady-state release Begin (yr)	End (yr)	Rate (Ci yr^{-1})
A	Spent fuel (PWR) 1 bundle	550	Crushable	Backfill	Sand/Bentonite	52·00	310	$8 \cdot 6 \times 10^5$	$8 \cdot 6 \times 10^6$	$2 \cdot 2 \times 10^{10}$	$1 \cdot 6 \times 10^{-9}$	1080	8000	$1 \cdot 9 \times 10^6$	$3 \cdot 9 \times 10^{-8}$
				Overpack	Iron	13·00									
				Canister	Mild steel	0·25									
				Stabilizer	Air	—									
B	Spent fuel (PWR) 1 bundle	550	Rigid	Backfill	Sand/Bentonite	13·00	1400	$3 \cdot 4 \times 10^4$	$3 \cdot 3 \times 10^5$	$1 \cdot 8 \times 10^{10}$	$1 \cdot 3 \times 10^{-12}$	1420	1690	$1 \cdot 8 \times 10^6$	$6 \cdot 7 \times 10^{-8}$
				Overpack	Ticode-12	0·25									
				Canister	Mild steel	0·25									
				Stabilizer	Cast lead	—									
A	Commercial HLW glass	2160	Crushable	Backfill	Sand/Bentonite	52·00	200	$8 \cdot 7 \times 10^5$	$8 \cdot 7 \times 10^6$	$4 \cdot 7 \times 10^9$	$4 \cdot 8 \times 10^{-13}$	900	8000	$2 \cdot 5 \times 10^6$	24×10^{-8}
				Overpack	Iron	13·00									
				Canister	Mild steel	0·25									
B	Commercial HLW grass	2160	Rigid	Backfill	Sand/Bentonite	13·00	530	$3 \cdot 5 \times 10^4$	$3 \cdot 5 \times 10^5$	$3 \cdot 1 \times 10^9$	$8 \cdot 1 \times 10^{-13}$	560	840	$2 \cdot 3 \times 10^6$	$4 \cdot 0 \times 10^{-8}$
				Overpack	Ticode-12	0·25									
				Canister	Mild steel	0·25									

Table 5.8 Performance of waste packages in hard-rock repository. Repository temperature $=165$ °C; hydrostatic pressure $=450$ p.s.i.; anoxic conditions [118].

Package type	Waste form	Heat loading (watts)	Crushable or rigid	Barrier element	Material	Thickness (in)	Time to waste-water contact (yr)	^{238}U release Breakthrough time (yr)	Begin (yr)	End (yr)	Rate (Ci yr^{-1})	^{99}Tc release Breakthrough time (yr)	Begin (yr)	End (yr)	Rate (Ci yr^{-1})
A	Spent fuel (PWR) 1 bundle	550	Crushable	Backfill Overpack Canister Stabilizer	Sand/Bentonite Iron Mild steel Air	52·00 13·00 0·25	1670	$1 \cdot 2 \times 10^6$	$1 \cdot 2 \times 10^7$	$2 \cdot 2 \times 10^{10}$	$7 \cdot 8 \times 10^{-13}$	2450	$9 \cdot 4 \times 10^3$	$1 \cdot 9 \times 10^6$	$3 \cdot 9 \times 10^{-8}$
B	Spent fuel (PWR) 1 bundle	550	Rigid	Backfill Overpack Canister Stabilizer	Sand/Bentonite Ticode-12 Mild steel Cast lead	13·00 0·25 0·25	1890	4850	$4 \cdot 7 \times 10^5$	$1 \cdot 8 \times 10^{10}$	$1 \cdot 3 \times 10^{-12}$	1900	$2 \cdot 2 \times 10^3$	$1 \cdot 8 \times 10^6$	$6 \cdot 7 \times 10^{-8}$
A	Commercial HLW glass	2160	Crushable	Backfill Overpack Canister	Sand/Bentonite Iron Mild steel	52·00 13·00 0·25	1630	$1 \cdot 2 \times 10^6$	$1 \cdot 2 \times 10^7$	$4 \cdot 6 \times 10^9$	$4 \cdot 8 \times 10^{-13}$	2410	9430	$2 \cdot 5 \times 10^6$	$2 \cdot 4 \times 10^{-8}$
B	Commercial HLW glass	2160	Rigid	Backfill Overpack Canister	Sand/Bentonite Ticode-12 Mild steel	13·00 0·25 0·25	530	5×10^4	$4 \cdot 9 \times 10^5$	$3 \cdot 1 \times 10^9$	$8 \cdot 1 \times 10^{-13}$	560	840	$2 \cdot 3 \times 10^6$	$4 \cdot 0 \times 10^{-8}$

and the corrosion resistance of the titanium-alloy case. In the salt repository the 'B' package performs better than the 'A' package because the combination of high corrosion resistance and rigid waste delay crushing and water entry. The 'A' package, on the other hand, is rapidly corroded and weakened to the point where it is crushed by the prevailing hydrostatic pressure. The situation in the hardrock repository is essentially similar but the corrosion rates are less than in the salt. While these canister types and designs may not necessarily be adopted in preference to any other concept they do provide a feel for the way waste canisters might be expected to behave in earth materials.

As a result of experimental work such as that undertaken by Lester [118–120] the most likely materials for the construction of waste canisters are titanium- and nickel-based alloys which have been shown to give low uniform corrosion rates. Cast iron, which has relatively much higher corrosion rates, may be applicable if sufficiently thick containers are suitable. Zircaloy has also demonstrated low corrosion rates but is generally discounted on cost terms, being markedly more expensive than titanium- and nickel-based alloys. Stainless steel, while having a good resistance to corrosion, is prone to stress cracking.

Spent fuel does not lend itself to vitrification but a number of feasible packaging options have been considered. In Sweden a molten lead filler or hot pressed copper powder is under review to embed the fuel rods in a solid matrix within the canister [113]. As with vitrified waste a filler material in the space between fuel pins and the container can enhance heat transfer, provide support against lithostatic pressure, act as a corrosion-resistant barrier, retard water intrusion, increase radiation shielding and enhance radionuclide sorption. The lower heat generation and much lower specific activities of α-bearing wastes do not require such rigorous packaging as spent and reprocessed fuel. Immobilization in concrete, bitumens or plastic resins followed by containment in steel drums and concrete casks will probably be all that is required [121].

5.5 BUFFER AND BACKFILL MATERIALS

Following through the multi-barrier approach many waste disposal models envisage the emplacement of a buffer or backfill material between the waste canister and the sides of the repository, which in many cases would be the rock into which the repository has been drilled or mined. The main functions envisaged for the backfill and buffer material are given in Table 5.9. For a detailed discussion of the role of backfill and buffer materials see Klinsberg and Duguid [122].

Table 5.9 The role of buffer and backfill materials.

To exclude water and prevent it from reaching the waste canister
To adsorb radionuclides which may escape from the waste canister
To buffer the groundwaters in the vicinity of the waste canister
To provide mechanical support to the canister and absorb stresses induced by
    possible rock movement
To dissipate the heat generated by the waste to the surrounding rock

The materials most commonly associated with engineered backfill, as it is sometimes known, are clays, sands and shales with various admixtures to increase the sorption potential. Nowak [119] gives a list of possible backfill and buffer materials and this is reproduced with minor modifications in Table 5.10. The effectiveness of the various backfill materials will depend to a large extent upon their chemical and physical stability over long periods. Pusch [123, 124] and other workers [119, 125–129] have carried out extensive tests on the suitability of various backfill materials which confirms the considerable chemical and physical stability of many of them, as

Table 5.10 Preliminary list of candidate backfill materials [119].

Clays
    Sodium bentonite
    Calcium bentonite
    Illite
    Treated soium bentonite
Sand
    Quartz sand (10–230 mesh)
Zeolites
    Clinoptilolite
    13X
    Zelon-900
Metal powders or fibres
    Iron
    Aluminium
    Lead oxide
Minerals/Rocks
    Pyrite
    Ferrosand (glauconite)
    Basalt
    Tuff
    Serpentine
    Anhydrite
Charcoal
Desiccants
    MgO and CaO

evidenced by their retention largely unchanged in naturally occurring rocks of considerable age.

The potential effectiveness of engineered backfill has been underlined by Nowak [128] who has shown that under appropriate conditions a thickness of 30 cm or so of suitable material can delay the release of ^{90}Sr and ^{137}Cs from the backfill for up to 1000 years. Most of the main fission products associated with high-level nuclear waste can be shown to be sorbed by either bentonite clay or zeolites [126] (see Table 4.2 for the exchange capacities of these and other potential backfill/buffer materials). By taking a mixture of materials, workers in the field hope to be able to design a backfill recipe for particular wastes and repository conditions. For instance, while the clay minerals in various bentonite mixes are reasonable for retaining Cs and Rb they are inadequate with respect to Sr, while non-synthetic zeolites on the other hand have been shown to be quite good at retaining Sr. Bentonite has the ability to swell when wet and this property could be used to good advantage in sealing microfractures in the rock immediately adjacent to the backfill. As water comes into contact with the bentonite it starts to expand outwards into the cracks through which the groundwater has arrived, partially sealing them in so doing, with the result that the further inwards migration of water is significantly reduced. The addition of dessicants such as CaO and MgO can also be included to reduce the volume of fluids by the production of $Ca(OH)_2$ and $Mg(OH)_2$, especially in evaporite repositories, for example [130]. The addition of quartz sand can be used to increase the strength of the backfill as well as improving its thermal conductivity. Thermal conductivity can also be enhanced by including a percentage of graphite and charcoal which have the additional advantage of being more selective in the sorption of Tc and I.

Clearly the final mix of any particular engineered backfill is likely to be a compromise involving a trade-off between physical and chemical properties of the various constituents. While bentonite, which is largely composed of montmorillonite clay, has good sorption and water take-up capacity (see Fig. 4.1) compared, for instance, with quartz sand, it has relatively much lower thermal conductivity and bearing capacity. Coons, Moore, Smith and Kaser [131] have overcome this problem in part by envisaging the use of two separate concentric layers about the waste to be deposited in a basalt repository. While an outer layer composed predominantly of bentonite is designed to prevent the ingress of groundwater the inner layer composed of more bentonite, other clays, quartz, phosphates, zeolites, basalt, sodium feldspar and bornite, amongst other materials, is directed primarily at the retention of radionuclides.

Where buffer and backfill materials are used it is clear they will be subject to the highest temperatures likely to be found in the repository and they will have an important role to play in helping heat move away from the repository into the surrounding rock. As Table 5.11 shows the thermal conductivity of buffer and backfill materials is significantly affected by moisture content and density. Temperature can also be important in terms of the long-term function of bentonite buffers. Dehydration of bentonites at about 390°C results in irreversible loss of the capacity of sodium bentonite clays to swell while more gradual effects may be expected at temperatures above 100°C.

Table 5.11 Thermal conductivity of candidate buffer/backfills [125].

Mix composition*	Moisture content range (5)	Dry density range (kg m^{-1})	Thermal conductivity in moist conditions W m^{-1} °C^{-1}	Thermal conductivity in dry conditions W m^{-1} °C^{-1}
WSAB-15	6–17	1100–1770	1·3–2·8	0·7–1·0
WSAB-25	5–16	1200–1760	1·2–2·7	0·6–0·9
WSAB-50	15–29	1330–1400	1·2–2·0	0·5–0·6
GSBB-10	6–10	1850–2000	2·0–3·0	0·4–0·7
GSBB-25	8–12	1500–1700	1·2–1·7	0·7–0·8
GSBB-50	12–20	1490–1520	0·7–1·2	0·5–0·7
GSSB-15	4–9	2050–2175	3·0–4·0	1·5–1·7
GSK-15	4–9	2000–2200	2·5–3·5	1·8–2·0
CGBB-15	5–19	1750–2000	1·3–2·6	0·8–1·0
CGBB-25	6–12	1620–1710	1·0–1·5	0·6–0·8
CGBB-50	12–17	1500–1550	0·8–1·0	0·5–0·6
CGBB-100	17–20	1320–1400	0·7–1·1	0·5–0·6
CGSB-15	6–10	1890–2200	2·4–3·4	1·4–1·8
CGSB-25	7–12	1600–2050	1·4–2·4	0·8–1·0
CGSB-50	9–15	1850–1980	1·8–2·4	0·9–1·0
CGSB-100				
CGK-15	6–10	1900–2050	2·2–3·0	1·7–2·0
CGK-25	9–12	1850–1980	1·7–2·5	0·9–1·2
CGK-50	9–15	1770–1860	2·1–2·2	1·0–1·2

*WS = Wedron sand; GS = Graded silica; CG = Crushed granite; AB = Avongel bentonite; BB = Blackhill (Wyoming) bentonite; SB = Sealbond; K = Kaolin; Z = Zeolite.

5.6 CONCLUSIONS

Many of the individual elements found in HLW are readily soluble in most groundwaters. While relatively dry rock repositories can probably be identified, transportation of radionuclides by groundwater remains the most likely threat to the overall integrity of geological containment. The solidification of HLW can greatly reduce the ability of groundwater to take HLW components into solution, though it cannot prevent it completely. The ingress of groundwater into the HLW can be considerably delayed by the use of appropriate waste canisters. Waste-canister design may eventually be capable of providing absolute isolation for the HLW from groundwater for hundreds of thousands of years. Radionuclides which are, however, mobilized in groundwater can have their rate of migration away from the repository markedly reduced by adsorption and other rock–groundwater interactions.

6 Repository options, design and construction

6.1 INTRODUCTION

Three main concepts are being considered for repository design: mined tunnel systems with waste emplacement in appropriately spaced drillholes, mined tunnels with waste emplaced in the tunnels themselves, and deep drillholes with waste emplaced directly from the surface. In general the concept of geological disposal would require the excavation of a void in the rock to provide access and room for waste emplacement. Provision may also be required to ensure adequate ventilation and safety for personnel who may have to work in the repository. The necessity to backfill and seal the repository also needs to be taken into account in the engineering design. As the previous chapters have shown different rock types have different hydrogeological, thermal and strength characteristics. In consequence any particular repository design will to a large extent be site- and host-rock-specific.

6.2 REPOSITORY SITE SELECTION GUIDELINES

The main features of interest at a repository site will be its chemical, physical and hydrological characteristics. These will need to be assessed in order to evaluate

'the induced stresses that might occur as a result of physical, chemical, thermal and radiation effects in the development of the repository and emplacement of the waste, the potential release mechanisms for waste radionuclides from the repository, and the pathways to the human environment, the potential for engineered features which could aid to isolate wastes in the case of unforeseen natural processes or events' [4].

It is important in obtaining the necessary data that the drilling of boreholes and other exploratory activity does not prejudice the future integrity of the repository.

Table 6.1 details the main factors to be examined at any specific repository site. Numerous site-specific design studies have now been carried out and they have demonstrated that it is possible to construct repositories in salt, mud rocks and crystalline rock using existing drilling, mining and civil engineering techniques. It remains to

Table 6.1 Factors to be examined at a specific repository site [4].

A. Characterization of the chemical, physical, geological, geomechanical and hydrogeological parameters

   1. Geological formation thickness and depth, lithology, mineralogy and stratigraphy
   2. Geological environment (including topography, structure, geological history, geomorphology, tectonics and seismicity)
   3. Hydrogeological environment (hydrological regime, including aquifers and aquicludes, areas of recharge and discharge)
   4. Geochemical environment (including water and rock chemistry, secondary minerals and sorption properties)
   5. Geotechnical and general physical properties of the host rock

B. Characterization of the natural surface environment

   1. Climate (normal and disruptive events such as floods, tornadoes, etc.)
   2. Surface hydrology
   3. Background radiation
   4. Flora and fauna

C. Characterization of the geological stability and resistance to:

   1. Climatic changes (glaciation, pluvial cycles, changes in sea level)
   2. Geomorphic activity (erosion, seismic activity, faulting, uplift or subsidence, volcanism)
   3. Meteorite impact

D. Characterization of the potential interactions between the waste and the host geological formation

   1. Thermal effects
   2. Radiation effects
   3. Hydrological effects
   4. Chemical reaction effects (sorption of radionuclides, corrosion of containers and conditioned waste, etc.)

E. Effects generated by man

   1. Pre-existing boreholes and shafts, excavation effects
   2. Likelihood of other man-induced events that could affect the safety of facilities (traffic density, fire, explosions, failure of man-made structures such as dams and sea-walls, etc.)

F. Economic and social considerations and planning

   1. Resource potential
   2. Land value and use
   3. Population distribution
   4. Jurisdiction and rights of land
   5. Accessibility and services
   6. Other environmental impacts
   7. Public attitudes

develop pilot repositories to provide actual field data before construction of full-sized operational repositories goes ahead; nevertheless, there is some justification in the NEA view that geological repositories can be built in practice [7].

6.3 THERMAL LOADING IN REPOSITORIES

As has been shown high-level nuclear waste can generate considerable quantities of heat over a number of years following reprocessing or its removal from the reactor. The choice of thermal loading for specific sites will involve multi-dimensional optimization that balances the effects of higher heat loadings in the short term and confidence in long-term safety against the desire to dispose of waste after only short periods of storage and cooling, or against the cost of building larger repositories if less waste is to be placed at any one location, to reduce the total heat loading.

The heat generated will depend not only upon the nature of the waste itself and its contained radionuclides, but also upon the width and length of the container, the decrease of heat generation with time, the thermal conductivity of the immobilized waste form, heat transfer to and through the waste packaging the buffer material and the surrounding rock, and the volume of waste. These variables are all closely interrelated. Owing to the variability between rock types and even within rock types it is not possible to stipulate a maximum design temperature for repositories generally. However it is possible to draw certain broad conclusions with respect to heat loadings.

Evaporites as we have seen have relatively high thermal values for heat conductivity. This will allow waste containers to be placed relatively close together whilst at the same time maintaining a homogeneous temperature field. The maximum temperatures in salt should not generally exceed 200°C [132] and a thermal load of $0.5–1$ W m^3 of salt, or about $1–2$ m^3 of salt per emplaced watt, together with a loading to 50–300 W m^{-2} in repository surface area, is likely to be acceptable [7].

In granites surface power densities in the range of 7–25 W m^{-2} have been proposed [132, 133]. In order to minimize the disturbance to the fracture distribution within such crystalline rocks it is desirable to keep temperatures as low as possible. While temperatures in the range 100–200°C are probably low enough to prevent the chemical alteration of buffer materials and to maintain the stability of the rock the Commission of the European Communities (CEC) studies suggest that a maximum temperature of about 20°C over the ambient on the full repository scale is appropriate in order to minimize the disturbance to the local geology and hydrology [132]. In crystalline rock it may also be desirable to design a wide flat repository in order to favour heat dissipation and minimize the potential for free convection in groundwater.

The situation with respect to clays and their relatively lower thermal conductivity values is very much site-specific. The NEA [7] illustrate this point by reference to the Mol clay repository site in Belgium where constraints on maximum temperatures at the clay/waste interface must be considered together with the requirement to limit the temperature rise at the interface of the clay and overlying aquiferous sands. As a general guide the CEC study [132] proposes a surface power density of 2.5 W m^{-2} for clay repositories.

Understanding of how heat moves through rock is quite good and calculated heat measurements have been shown to correspond reasonably well with field measurements carried out on a relatively large scale. At the Mol site clays a heater was buried at a maximum depth of 650 cm and surrounded by a total of 36 thermoprobes at various distances from the heat sources. The experiment ran for one year at 1000 W(e) output and for another six months at 1500 W(e) [89]. These experiments showed a very high degree of correlation between predicted and measured temperatures and gave an average *in situ* value for heat conductivity of 1.69 W m^{-1} °C^{-1}, and 18.8 m^2 yr^{-1} for heat diffusivity. Preliminary heater experiments have also been carried out on the Climax granite stock of the Nevada test site [122]. These experiments have shown that the thermal conductivities measured in the field are on average about 10 to 20% higher than laboratory-obtained values. Equally interesting, these experiments suggest that there are marked differences between the transfer of heat across fractures than parallel to fractures. Thermal conductivity values across fractures are about 10% higher than parallel to the fractures. This work also suggests that rock permeability decreases with rising temperature. This may be due to an increased amount of fracture annealing at higher temperatures.

6.4 SUBSURFACE EXCAVATION

The successful development of subsurface excavations is to a large extent dependent upon the level of knowledge of the geology to be encountered. A picture of the subsurface geology can be built up using geophysical methods and boreholes and appropriate downhole logging techniques. Seismic refraction can give information on strata distribution, fault locations, weathered zones and rock quality. Resistivity techniques can also aid initial investigations providing information on water tables and saturated formations. Ultimately the construction of pilot tunnels will yield the most valuable information prior to the construction of the full-size repository [134].

Two main methods are currently used for the construction of underground tunnels and caverns. In harder rocks blasting is required to loosen the rock whilst in softer formations continuous boring machines may be used. The main advantage of using the boring machine is that it tends to give a smoother finish to the excavated void and, perhaps more importantly, it limits the fracturing of the rock surrounding the excavations compared with blasting techniques [135]. On this point it is worthy of note that new pre-split blasting techniques have demonstrated in the Stripa project in Swedish granite that the zone of disturbed permeability can be limited to about one metre from the excavation walls. These results have been confirmed by independent work carried out by the Colorado School of Mines [136]. The use of tunnelling machines also reduces the degree of overbreak from as much as 25% for blasting methods to about 5%. Tunnelling machines have successfully cut most sedimentary rock types but their use on the harder igneous and metamorphic rocks is still limited. Muirhead and Glossop [137] suggest that the present economic limit of using tunnelling machines is confined to those rocks with an unconfined compressive strength of $200 \, \text{MN m}^{-2}$. Drivage has been achieved in rocks with compressive strengths of up to $300 \, \text{MN m}^{-2}$ but this would not generally be economic owing to the excessive wear and abrasion of the cutting surfaces. Pirie [138] has produced a graph showing the reduction of machine excavation rate with increasing compressive strength of rock; this is reproduced as Fig. 6.1.

Where blasting techniques are used to drive headings a number of different methods are available as shown in Fig. 6.2. The most

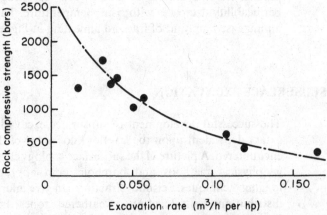

Figure 6.1 Curve showing reduction of machine excavation rate with increasing compressive strength of rock [138].

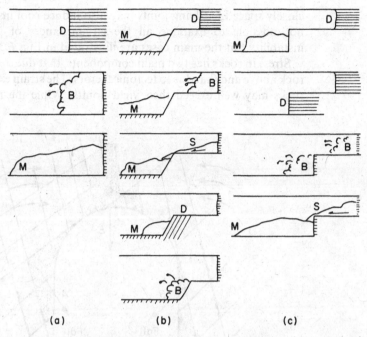

Figure 6.2 Tunnelling methods. A: full face; B: top heading and bench; C: Top heading, bench drilled horizontally. Phases. D: drilling; B: blasting; M: mucking; S: scraping [139].

common method of tunnelling in hard rock uses a full-face blasting pattern, with an entire section of the face extending from the floor to the roof being removed at each cut. The amount of explosives used varies between about 0.9 kg m^{-3} in large-diameter tunnels to about 3·6 kg m^{-2} in small-diameter drives [139]. The general principle when blasting is to produce a cavity in the centre of the face into which subsequent blasts can move. The use of a number of small charges detonated with microsecond delays is preferable to a single charge of equivalent weight as it reduces the vibration levels and peak particle velocities and limits the zone of rock disturbance around the excavation.

As noted earlier relatively smooth cavity profiles can be achieved using the pre-splitting technique. This technique uses a low charging density in closely spaced perimeter holes. These perimeter charges are detonated prior to the main charge. The effect of this is to limit crack formation to the zone between the drill holes and to confine the cut within the required cavity contours.

To a large extent overbreak and tunnel stability will be determined by joint patterns within the rock. Where jointed strata dip steeply into cavities at 30° or more the up side may be unstable. The presence of closely spaced flat-lying joints may also induce roof instability. It is not possible to examine all the circumstances of joint-induced instability but the main cases are illustrated in Figs 6.3 and 6.4.

Stress in rocks has two main components, that due to the weight of rock above and that due to tectonic factors. The strain experienced by rocks may well exceed their yield points. While the rocks remain

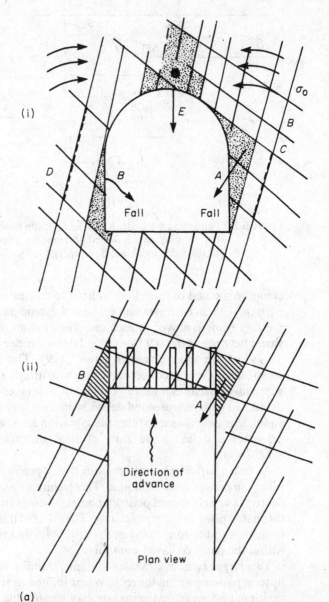

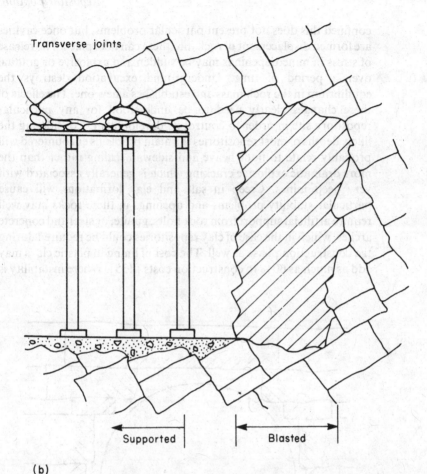

Transverse joints

Supported | Blasted

(b)

Figure 6.3 (a) Tunnel in rock with steeply dipping joints. (i) Steeply dipping joints (45–90°) which are parallel to the tunnel axis lead to slabbing of the wall and fallouts from the roof. At point A the slab 'daylights' at the feather-edge bottom and would probably fall with the force of the blast during tunnel advance. The slab B may not fall, however; it could be loosened by the original blast and would be susceptible to additional loosening by the shocks of later blasts and by the 'working' of the rock under peak tangential stresses around the tunnel periphery. Unless restrained, the slab B might eventually fall. Joints at depth such as C and D may tend to open. Joint blocks at E may be extremely dangerous, appearing stable after the blast but becoming unstable as the tunnel advances and the rock adjusts to the new stress field. (ii) Block A may be loosened and possibly forcefully ejected by gas pressures during blasting. Block B might be loosened but not necessarily removed. Had the tunnel advanced in the opposite direction the relative positions of A and B would be interchanged. (b) Tunnel driven perpendicular to the strike of steeply dipping and jointed rocks [139].

confined this does not present particular problems, but once cavities are formed displacement of rock into these can take place. The release of stress in mined openings may be sudden and explosive or gradual over a period of time. Underground excavation destroys the equilibrium in the rock mass and establishes a new one. The effects of these changes clearly need to be understood for any particular repository site in order to control its development. Considering the likely depths of most repositories the main problems encountered will probably relate to floor heave and sidewall scaling rather than the more dramatic explosive cracking which is generally associated with very deep mines. Creep in salt and clay formations will cause particular stability problems and openings in these rocks may well require artificial support from rock bolts, gunite, or steel and concrete arches, which in the case of clay repositories could be required during the construction phase as well. The cost of lining in plastic clays may add as much as 60% to construction costs [135]. Where instability is

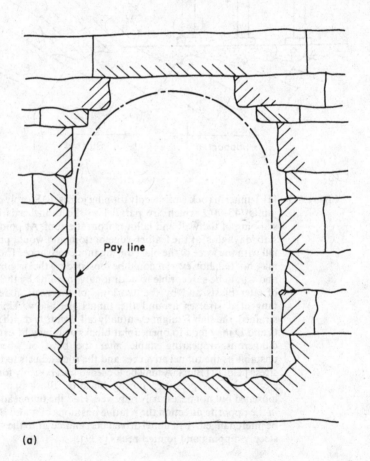

Pay line

(a)

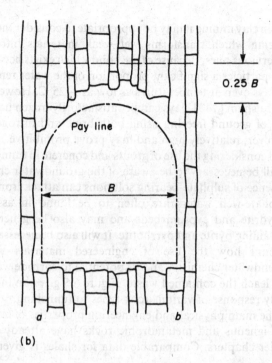

(b)

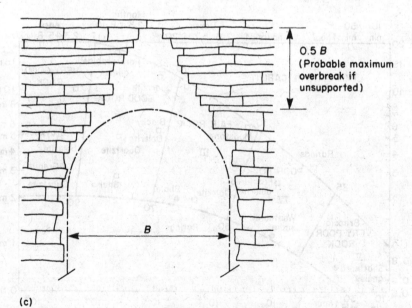

(c)

Figure 6.4 (a) Bridge action in strong rocks with few joints. (b) Axis of tunnel running parallel with strike in vertically dipping rocks. (c) Overbreak in thinly bedded horizontal strata with joints. Ultimate overbreak occurs if no support is installed [139].

met in clay mining it may be appropriate to consider the use of ground freezing which transforms wet weak materials into free-standing impervious ones. Because of the bond which exists between water and clay particles a significant proportion of the water remains unfrozen in clays even at temperatures as low as −25°C. However, according to Maisham [140] clays can be satisfactorily frozen nonetheless. The cost of ground freezing using both brine and nitrogen coolants is, however, relatively high and may prove prohibitive.

In considering the use of grouts and concrete in tunnels and cavities it will be necessary to be aware of the groundwater chemistry as the presence of sulphate-bearing solutions can attack iron and concrete. Sulphate-rich waters are often to be found in association with anhydrite and gypsum beds and may also be generated by rocks containing pyrite and pyrrhotite. It will also be necessary to take into account how the use of engineered materials will affect the groundwater chemistry and its potential to corrode waste containers and leach the contained waste. Figure 6.5 gives an indication of the likely response of various rock types to tunnelling.

The main physical and engineering properties of evaporites, clays and igneous and metamorphic rocks have already been given in earlier chapters. Comparable data for shales is given in Table 6.2.

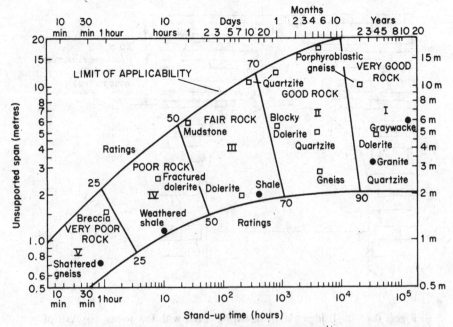

Figure 6.5 Geomechanics classification of rock masses for tunnelling. South African case studies are indicated by squares while those from Alpine countries are shown by dots [139].

Table 6.2 Engineering evaluation of shales (modified from [40]).

Physical properties	Average range of values		Probable in situ behaviour			
Laboratory tests and in situ observations	Unfavourable	Favourable	High pore pressure	Low bearing capacity	Tendency to rebound	Slope stability problems
Compressive strength (kPa)	350–2070	2070–3500		✓		✓
Modulus of elasticity (MPa)	140–1400	1400–14000			✓	✓
Cohesive strength (kPa)	35–700	700–>10500				✓
Angle of internal friction (degrees)	10–20	20–65				✓
Dry density (Mg m⁻³)	1·12–1·78	1·78–2·56			✓	
Potential swell (%)	3–15	1–3	✓			✓
Natural moisture content (%)	20–35	5–15	✓			✓
Coefficient of permeability (m s⁻¹)	10⁻⁷–10⁻¹²	>10⁻⁷	✓			

While it is unlikely that repositories would be constructed in sandstones, coal measures or limestones it may be necessary to mine through such rock types to reach more appropriate rock environments. In this regard brief summaries of the engineering properties of these rock types are given in Tables 6.3–6.5.

6.5 GEOTHERMAL GRADIENTS

The rate of increase in rock temperature with depth is dependent upon the geothermal gradient which averages an increase of 1°C for each 30 to 35 m of depth. The actual rate of temperature rise with depth will vary considerably from site to site. According to Bell [139] in volcanic areas it may be as high as 1°C for each 10 to 15 m, whilst in more geologically stable areas it may be as low as 1°C for each 60 to 80 m. The geothermal gradient below mountains is also generally less than that below valleys. There would appear to be advantages in choosing sites with relatively low geothermal gradients so allowing for the disposal of a greater heat load at any one repository. An indication of the variability in geothermal gradients can be gained from Table 6.6 [139].

6.6 EXAMPLES OF REPOSITORIES

6.6.1 THE BASALT WASTE ISOLATION PLANT AT THE HANFORD SITE, UNITED STATES OF AMERICA

This is a site located in the midst of the Columbia River Basalts which were formed about 15 million years ago. The flows of most interest for waste emplacement lie at a depth of between 800 and 1400 m. The mineralogy of the basalts which are for the most part fine-grained is dominated by a ground mass of plagioclase, augite, ilmenite, pigeonite, titanomagnetite, olivine and some orthopyroxene. They also contain microphenocrysts of plagioclase and augite. Fractures occur in the basalts, ranging in size from about 100 μm to 1 cm in width. These fractures are frequently filled with clay minerals, which not only have an important role to play in restricting the fracture porosity and permeability of the rock but are potentially powerful sorption materials. The various lava flows found in the area show a high degree of uniformity but brecciated horizons do occur. Surprisingly, according to Brookins [10] these breccias with their markedly higher porosities and permeabilities have not allowed water to flow to underlying basalts. Nevertheless, the flow contacts at the boundaries of different lavas possess relatively high values for

Table 6.3 Some physical properties of arenaceous sedimentary rocks [40].

	Fell sandstone	Chatsworth grit	Bunter sandstone	Keuper waterstones	Horton flags	Bronllwyn grit
Relative density	2·69	2·69	2·68	2·73	2·70	2·71
Dry density ($\times$Mg m^{-3})	2·25	2·11	1·87	2·26	2·62	2·63
Porosity	9·8	14·6	25·7	10·1	2·9	1·8
Dry unconfined compressive strength (MPa)	74·1	39·2	11·6	42·0	194·8	197·5
Saturated unconfined compressive strength (MPa)	52·8	24·3	4·8	28·6	179·6	190·7
Point load strength (MPa)	4·4	2·2	0·7	2·3	10·1	7·4
Scleroscope hardness	42	34	18	28	67	88
Schmidt hardness	37	28	10	21	62	54
Young's modulus ($\times 10^3$ MPa)	32·7	25·8	6·4	21·3	67·4	51·1
Permeability ($\times 10^{-9}$ m s^{-1})	1740	1960	3500	22·4	—	—

Table 6.4 Some physical properties of carbonate rocks [40].

	Carboniferous limestone	Magnesian limestone	Ancaster freestone	Bath stone	Middle chalk	Upper chalk
Relative density	2·71	2·83	3·70	2·71	2·70	2·69
Dry density (Mg m^{-3})	2·58	2·51	2·27	2·30	2·16	1·49
Porosity (%)	2·9	10·4	14·1	15·6	19·8	41·7
Dry unconfined compressive strength (MPa)	106·2	54·6	28·4	15·6	27·2	5·5
Saturated unconfined compressive strength (MPa)	83·9	36·6	16·8	9·3	12·3	1·7
Point load strength (MPa)	3·5	2·7	1·9	0·9	0·4	—
Scleroscope hardness	53	43	38	23	17	6
Schmidt hardness	51	35	30	15	20	9
Young's modulus ($\times 10^3$ MPa)	66·9	41·3	19·5	16·1	30·0	4·4
Permeability (-10^{-9} m s^{-1})	0·3	40·9	125·4	160·5	1·4	13·9

Table 6.5 Engineering properties of some Coal Measures rocks [40].

	Mudstone	Siltstone	Shale	Barnsley Hards coal	Deep Duffryn coal
Relative density	2·69	2·67	2·71	1·5	1·2
Dry density (Mg m^{-3})	2·32	2·43	2·35	—	—
Dry unconfined compressive strength (MPa)	45·5	83·1	20·2	54·0	18·1
Saturated unconfined compressive strength (MPa)	21·3	64·8	—	—	—
Point load strength (MPa)	3·8	6·2	—	4·1	0·9
Scleroscope hardness	32	49	—	—	—
Schmidt hardness	27	39	—	—	—
Young's modulus ($\times 10^3$ MPa)	25	45	5·2	26·5	—

Table 6.6 Gradients for some European tunnels [40].

Tunnel	Length (m)	Depth (m)	Average geothermal gradient (m °C^{-1})	Maximum temperature (°C)
Simplon	19 720	2135	37	55
St. Gotthard	14 998	1752	47	40
Mont Cenis	12 236	1610	58	30
Tauern	8 551	1567	24	49

hydraulic conductivity of between 10^{-6} and 10^{-9} m s^{-1}, with porosities of a few percent. The confining basalts in contrast have a hydraulic conductivity estimated to be in the order of 10^{-10} m s^{-1} or less, and estimated porosities of 1–2%. Hydraulic head measurements indicate a gradient of about 1.0 to 0.1 m km^{-1}, with flow being predominantly horizontal [141]. Benson, Carnahan, Apps [76] have given data relating to the composition of groundwaters in the basalts and these are shown in Table 3.8.

Smith, Turner and Deju [142] have described how these basalts might be used as a repository, taking high-level waste equivalent to approximately 52 000 tonnes of heavy metal, assumed to have been stored for 10 years before disposal. Capacity is also provided for 32 000 drums of α waste. The emplacement rooms, designed to accommodate maximum thermomechanical rock stresses, would be 3 m high and 6 m wide with arched roofs. High-level waste would be placed in horizontal holes of 70 cm diameter, extending up to 60 m from room sidewalls. With this design retrieval would be possible for a limited period. An artist's impression of the repository is given in Fig. 6.6 [7].

6.6.2 A CONCEPTUAL DESIGN FOR A REPOSITORY IN THE BOOM CLAY FORMATION AT MOL IN BELGIUM

In Belgium the only rocks considered suitable for HLW disposal are clays and shales. Seven potential areas have been identified, one of which is located in the north-eastern part of the country above the Boom Clay Formation. A cored borehole was drilled to the top of the Cretaceous some 570 m below ground level. The Tertiary Boom clay was met between approximately 160 and 270 m below ground level. The Boom clay is a rather homogeneous, stiff but still plastic clay, containing up to 25% of water, about 20% of smectite and a variable percentage of carbonaceous material [89]. Laboratory experiments on clay cores indicate a thermal conductivity of 1·5 W m^{-1} °C for

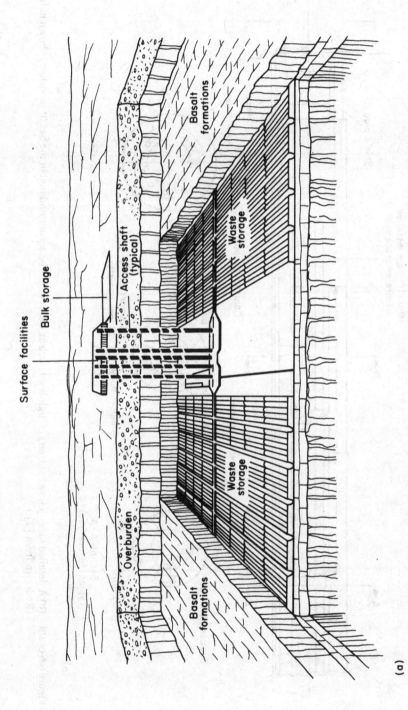

Figure 6.6 (a) An artist's impression of a basalt high-level nuclear waste repository.

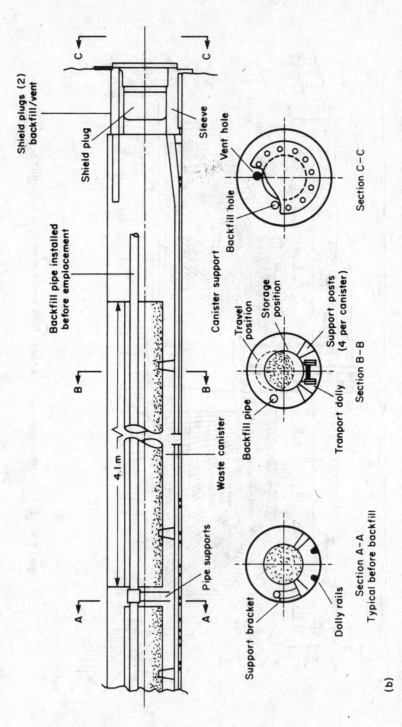

Figure 6.6 contd (b) A detailed view of how canisters of high-level nuclear waste might be accommodated in the basalt repository shown in (a). (Redrawn from [7].)

(b)

natural clay containing up to 25% water and at an average temperature of around 35°C. This value of thermal conductivity, however, was noted to fall very rapidly upon heating of the clay and at 100°C it had reduced to 0·3 W m^{-1} °C. Water convection in the clays is thought to be negligible even under repository conditions [89]. With respect to the retardation of radionuclides, the presence of vermiculite and smectite clay minerals makes the Boom clay an excellent sorbent with a measured cation exchange capacity of 0·3 ± 0·05 meq g^{-1}.

It is envisaged that a repository in the Boom clay would accommodate some 9000 heat-generating high-level waste containers each 0·3 m in diameter and 1·5 m in length and holding 100 litres. The waste would derive from a 30 year 10 Gw(e) nuclear power programme. It is also thought that a further 9000 containers of non-heat-generating wastes might be emplaced, together with 150 000 medium-level α-waste containers of 220 litres, and 6000 220-litre α-waste containers. The high-level waste would be vitrified and stored for 50 to 75 years before disposal in order to reduce its heat output significantly. Before disposal fuel cladding waste would be compacted and embedded in a metal matrix and α wastes would be incorporated in bitumen or another acceptable medium.

The repository would consist of a series of horizontal galleries of circular cross section, 3.5 m in diameter and 1250 m long at a depth of about 200 m. Two access shafts are envisaged together with a 4.5 m diameter, 550 m long head gallery. Lined storage holes each 21 m long and spaced about 12 m apart would be bored from the main gallery at 45° from the vertical, and each would contain twelve containers of waste. This arrangement is thought to provide an optimum balance of disposal capacity and excavation costs within the constraints of thermal loading and vertical extent of the Boom clay. The repository is designed to a maximum thermal load of about 15 kW per hectare, with a maximum temperature elevation of about 85°C in the clay at 1 m from a canister, 5°C at the interface of overlying aquiferous sands and less than 0·5°C at the surface [7, 143]. Prospective views of the proposed repository are given in Fig. 6.7.

6.6.3 A DEEP DRILLHOLE DESIGN FOR HIGH-LEVEL WASTE EMPLACEMENT IN THE MORS SALT DOME IN DENMARK

Pedersen and Lindstrom–Jensen [38] have given a description of the geology of the Mors Salt Dome. The Mors Salt Dome is located beneath the island of Mors in the inlet of Limfjorden in Denmark. The salt dome extends to a depth of about 5·5 km below the surface and is approximately 8 km in diameter. Overlying the salt is between 650 and 900 m of chalk. Tertiary clays which occur elsewhere in this part

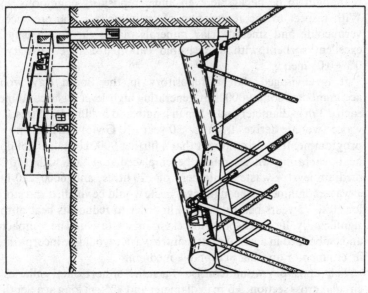

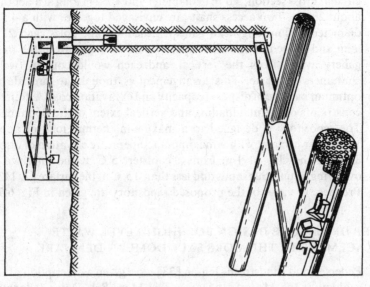

Figure 6.7 Perspective views of a design for the emplacement of intermediate-level waste drums (a) and high-level waste canisters (b), in the Boom Clay Formation at Mol. (Redrawn from [143].)

of Denmark have been removed by glacial action in the vicinity of Mors. The salt in the dome appears to be quite pure with halite accounting for 97–99% of the rock. Small amounts of anhydrite are also present. Towards the periphery of the dome, clays and carnallite have been recorded. It is probable that the salt has undergone considerable internal folding. The rock temperature at the top of the dome is about 26°C and increases at about 2°C per 100 m. The water content of the salt as measured from cores is generally very low, being below 0·01%, though small amounts of brine were recorded in cores taken from depths of between 1800 and 2000 m. Permeability in the overlying chalk is relatively low with estimated aquifer velocities of only 0·01 m y^{-1} apparently posing no surface dissolution problems. The matter of internal dissolution has also been considered with respect to the presence of brines, anhydrite and carnallite within the dome. Internal dissolution is not thought to pose problems. While brines are known to migrate in temperature fields, towards the higher temperatures if the inclusions are pure liquids, and away from them if they also contain a gas phase, the maximum calculated migration in this case is put at 30 m in one million years.

The original design requirement was for a waste repository corresponding to the operation of six 1000 MW(e) light-water reactors for 30 years. The waste is assumed to be vitrified and to undergo 30 years of storage before disposal. In order to avoid impurities noted in salt cores the holes would be concentrated at depths of between 1200 and 2500 m towards the centre of the dome as shown in Fig. 6.8. Temperatures within the salt are not expected to rise above 100°C. An overpack of steel some 15 cm thick would surround each waste canister which is sufficient to withstand lithostatic pressures at a depth of 2500 m. The repository design envisages the containment of 5200 canisters of 150 litres. These would be contained in eight holes spaced around a circle of radius 500 m. The holes would be lined to a depth of 950 m, below which an unlined hole of 750 mm diameter would be drilled to a total depth of 2500 m. This unlined section would allow salt creep to cover and seal the waste after closure.

The main advantage of the borehole concept is its relatively low cost. There is no need for mining and all drilling is done from the surface. Against this the volumetric capacity is limited and the method is only suitable for thick, homogeneous formations.

6.6.4 AN EXAMPLE OF THE USE OF AN OLD MINE WORKING – THE FORMER 'KONRAD' IRON-ORE MINE, FEDERAL REPUBLIC OF GERMANY

The 'Konrad' iron-ore mine was closed down in 1976 for economic

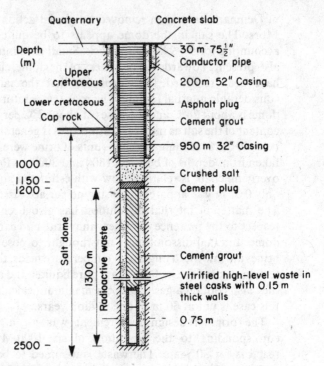

Figure 6.8 Schematic diagram of a deep-hole disposal facility.

reasons. However, the geology of the site as shown in Fig. 6.9 is such that the mine has been considered as a suitable site for the deposit of non-heat-generating nuclear wastes, being dominated by practically impermeable Albian, Neocomian and Jurassic claystone formations [144]. According to current planning a waste volume of 400 000 m^3 could be accommodated. This amount of waste would have an activity of approximately 10^{+17} Bq corresponding to a heat production of about 10^{+4} W. The basic storage proposal is to use large volume galleries with a cross section of 40 m^2 or so as shown in Fig. 6.10.

6.6.5 THE CANADIAN REFERENCE DESIGN FOR A HIGH-LEVEL WASTE REPOSITORY

In the Canadian concept of high-level nuclear waste disposal two baseline studies have been developed by Burgess and Sandstrom [145]. These envisage the containment of immobilized reactor waste in holes drilled in the floor of an underground chamber followed by immediate backfilling, and the emplacement of irradiated CANDU reactor fuel in the chambers. The irradiated fuel would be immediately surrounded by at least 1 m of backfill, with the remainder of the

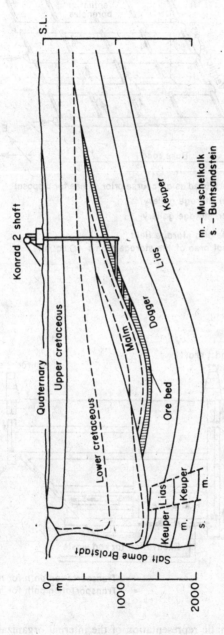

Figure 6.9 The geological setting of the Konrad iron ore mine. (Redrawn from [7].)

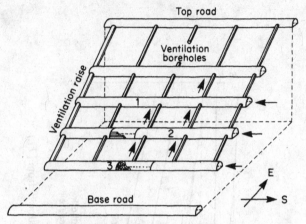

1. Gallery used as air course prior to use for disposal
2. Upper storage gallery
3. Lower storage gallery

Principle of a storage field
(sectional area of the storage rooms: 40 sq.m.)

(a)

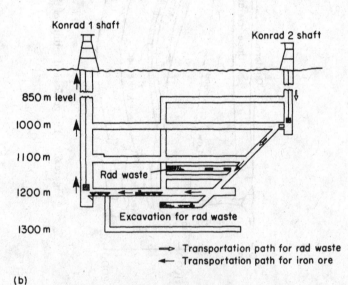

(b)

Figure 6.10 Schematic representation of the internal organization of the Konrad iron ore mine as it might be laid out to accommodate nuclear waste storage. (a) shows the ventilation arrangements and (b) the transport proposals. (Redrawn from [7].)

chamber being left open for 20 years or longer after emplacement to allow ventilation cooling before complete backfilling. The optimal distance between individual waste canisters is put at 1·5 m, with rows of irradiated fuel and immobilized waste being 2·5 m and 1·5 m apart respectively as shown in Fig. 6.11. It is envisaged that both types of waste would be stored for 10 years prior to disposal, and at the time of emplacement would have a thermal loading of 269 W per canister. The initial panel thermal loadings corresponding to the canister arrangements shown in Fig. 6.11 are 24 W m^{-2} for immobilized waste and 14 W m^{-2} for irradiated fuel. The *in situ* stress of the repository rock would be of the order of 48 MPa with a corresponding vertical stress of about 28 MPa. The repository itself would be sited at a depth of 1 km in a suitable pluton in the Canadian Shield. Assuming vertical drillholes were used, as opposed to horizontal ones, the temperature adjacent to the waste would reach a maximum of about 120°C, reducing to about 85°C after 100 years. Lower initial temperatures would be possible by using horizontal boreholes but after 100 years the difference is only 2°C as opposed to 20°C one year after disposal. These differences are due to the greater distance of the horizontally placed waste from the centreline of the waste chamber as shown in Fig. 6.12.

To gain actual field knowledge and experience the Canadians have embarked upon the construction of an underground research laboratory in a previously undisturbed granite batholith near the town of Lac du Bonnet in Manitoba, close to the Whiteshell Nuclear Research Establishment [146]. To date the laboratory has been excavated to a depth of 254 m, some 245 m below the water table, including a vertical access shaft, a ventilation bore and a main horizontal experimental level [147].

6.6.6 A TUFFACEOUS ROCK REPOSITORY

Sykes and Smith [78] have described a thick but variable sequence of tuffaceous rocks at the Nevada Test Site near Yucca Mountain in the United States of America. A geological cross section of the area is given in Fig. 6.13. The various units which make up the tuff sequence show considerable lateral and vertical variation. The oldest unit is a non-welded ash-flow tuff. Immediately above this is the Bullfrog Member, a welded ash-flow tuff in this case. Above this is a vitric crystal tuff which contains welded and non-welded units. Above these lie rocks showing yet more variation including air-fall, ash-flow tuffaceous rocks and volcanoclastic sedimentary units. Brookins [10] has highlighted the importance of the presence of zeolites in this 900 m + thick sequence of tuffaceous rocks. These minerals as noted earlier can play an important part in halting the migration of

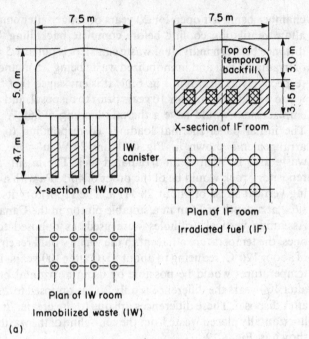

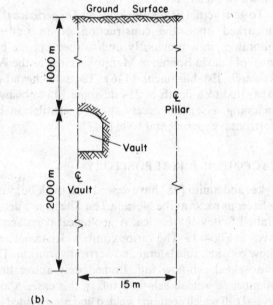

Figure 6.11 The Canadian reference design for a high-level nuclear waste repository in crystalline rock. (a) shows the geometry of the immobilized waste (IW) and irradiated fuel (IF) room and canister boreholes, while (b) shows the boundary and initial conditions used in subsequent thermal and stress analysis. (Redrawn from [58].)

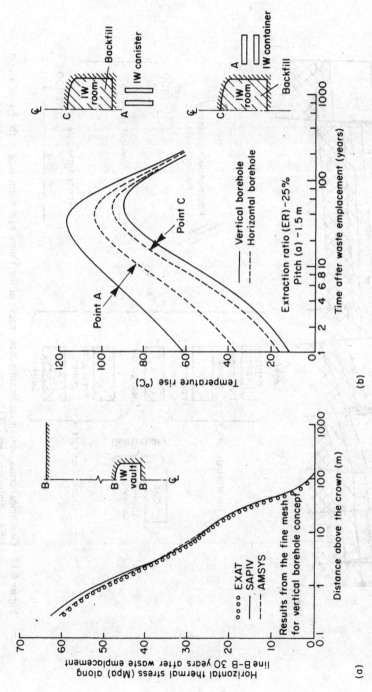

Figure 6.12 (a) Comparison of stresses obtained using three different computer codes. (b) Comparison of calculated temperatures around a repository using vertical- and horizontal-borehole concepts.

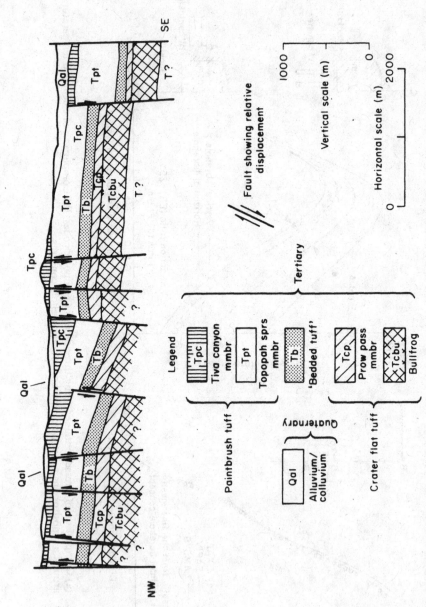

Figure 6.13 Geological cross section of the tuffaceous rocks of the Yucca Mountain area of the Nevada Test Site, USA. (Redrawn from [78].)

radionuclides owing to their relatively high cation exchange capacity. Brookins [10] points out that Johnstone and Wolfsberg [148] have produced a study of the sorption–desorption properties of the Nevada Test Site tuffs and some of their data are shown in Table 6.7. These data are reported as a sorption factor, R_d, in ml g^{-1}. As can be seen all the values are large with the exception of I and C. Brookins [10] also points out that where data have been recorded under both reducing and oxidizing conditions, the larger R_d values are reported for reducing conditions, which underlines the importance in the majority of cases of chemically reducing environments to retain or retard possible radionuclide movement.

The conceptual design for waste disposal in these tuffs assumes a repository located within the Bullfrog Member at a depth of 800 m below ground level [79]. The Bullfrog Member is composed primarily of devitrified welded tuff composed largely of quartz and

Table 6.7 Empirically derived R_d (ml g^{-1}) values from batch experiments at 25°C, 0·1 MPa for selected radionuclides in order of ranking [40].

Element	Salt	Basalt	Tuff	Granite
Tc	2*	20,† 0	10,† 0	4*
Pu	500,† 50	200,† 100	500,† 40	500,† 100
Np	30,† 7	50,† 3	50,† 3	50,† 1
I	0	0	0	0
U	1*	6*	4*	4*
Cs	1800	300	100	300
Ra	5	50	200	50
Sr	5	100	100	12
C	0	0	0	0
Am	300	50	50	200
Sn	1,† 50	10,† 100	50,† 500	10,† 500
Ni	6	50	50	10
Se	20, 100,† 20	20,† 5	2	2
Cm	300	50	50	200
Zr	500	500	500	500
Sm	50	50	50	100
Pd	3	50	50	10
Th	50, 100	500	500	500
Nb	50	100	100	100
Eu	50	50	50	100
Pa	50	100	100	100
Pb	2	25	25	5
Mo	0, 5,† 1	10,† 4	10,† 4	5,† 1

*No significant difference between value measured in oxidizing and reducing E_h's.
†Reducing conditions, second-value oxidizing conditions. First value for dome salt, second for bedded salt.

K-rich alkali feldspar with zeolite alteration products [149]. The ambient rock temperature is assumed to be 35°C. The gross thermal loading would be of the order of 25 W m^{-2}. Waste would be stored for ten years, followed by disposal in a series of long parallel rooms with two rows of waste canister emplacement holes drilled parallel to the room centreline. While the area is known to be extensively faulted the initial concept envisages the repository being confined within a fault-bounded block. The thermal conductivity of the tuff is put at 1·55 W m^{-1} °C when dry and 2·40 W m^{-1} °C when wet. The groundwater conditions are given in Table 3.11. While the boiling of groundwater is unlikely to cause problems [79], the presence of a relatively high potassium-to-sodium ratio (0·1) in the groundwater may affect the stability of Na-montmorillonite in a hydrothermal environment [150]. The use of bentonite as a backfill on closing the repository after about 50 years could also be threatened by the high temperatures generated in the repository; according to the assumptions made by Johnstone, Sundberg and Krumhansl [80] such temperatures could reach 535°C. At temperatures above 390°C dehydration of bentonite results in irreversible loss of its swelling capacity. The thermal loading could be reduced, however, by reducing the heat loading per acre. With a heat loading of 0·55 kW per canister and 50 kW per acre, the maximum temperature calculated by Johnstone, Sundberg and Krumhansl [80] was just over 100°C. With heat loadings of 1·6 kW per canister and 60 kW per acre this rises to just under 300°C.

6.6.7 DEEP-WELL CONCEPTS

A number of deep-well concepts have also been considered. They all envisage very deep disposal on the assumption that if waste is placed sufficiently far down in the Earth's crust it will be effectively isolated from the biosphere. As such, these methods rely not upon the multi-barrier principle but solely on the ability of the rock to isolate the waste. That is not to confuse such methods with the emplacement of waste canisters in deep repository boreholes.

One of the methods which has been suggested is to pump a slurry of reprocessed HLW down deep boreholes into underground caverns. After a few years when the water in the slurry has been driven off by the heat generated by the decaying waste, the remaining heat would reach sufficiently high temperatures to melt the surrounding rock. Assuming burial depths of about 2000 m rock melt and resolidification should be complete within about 1000 years. The waste would then be immobilized within the rock matrix and, it is hoped, stabilized. The disadvantages of such a scheme are many. The preparation of the slurries would not be easy, the movement of groundwater at these depths is not clearly understood and the removal of water from the hole during the dewatering phase of the slurry is problematical. The concept is also one where the waste is effectively non-retrievable if unforeseen problems arise.

A second deep-disposal method involves deep-well injection of liquid HLW into permeable and porous strata overlain by impermeable formations. While the most obvious rocks to examine in this context may be the sandstones and limestones which have already been well characterized by the oil and gas industries, of more immediate interest is the potential use of artificially fractured shales with their high cation exchange capacities. In this method the shale is first fractured by water injected under high pressure, a technique now well understood as a result of the hot dry rock geothermal programmes at Los Alamos in the United States and at Camborne in the United Kingdom. The waste is then injected together with a clay slurry and allowed to solidify in place. As the natural permeability of the shales is low the waste would not be expected to migrate under the influence of moving groundwater. While again this is a non-retrievable method of disposal it does have the advantage that considerable experience is already available with respect to the disposal of toxic liquid wastes into shale horizons in the United States. The Oak Ridge National Laboratory has placed nearly 7·5 million litres of ^{137}Cs- and ^{90}Sr-bearing wastes using the shale grout method [151].

6.7 BACKFILLING AND SEALING REPOSITORIES

The role and requirement for backfilling has already been outlined in Chapter 5. The extent to which backfilling is necessary will depend upon the repository concept employed and the characteristics of the host rock. Salt repositories and free-fall sub-seabed concepts may be entirely self-sealing. Crystalline rock repositories in comparison will require a fully engineered backfill and sealing operation. Certain plastic clays, on the other hand, may be partly self-sealing. The choice of backfill will be influenced by the nature of the host rock and the desire to produce a hydrologic regime which would limit flow paths to the biosphere. It will need to be compatible with the rock and the waste package and should have chemical and physical durability. Backfill and sealing materials have recently been reviewed by Butlin and Hills [152] and by Mott, Hay and Anderson [153].

For certain applications crushed or otherwise treated rock removed to create the repository may be acceptable. Such material enjoys a high degree of compatibility with the repository rock and should be readily available at relatively low cost. In general it is likely that such material would need to be crushed and mixed with clays, cements or fly ash to reduce its permeability and increase its sorption capacity. In the United States the use of melted or crushed salt and bentonite has been investigated with some success [154, 155]. There is also considerable experience available in the mining and construction industries on the use of clay and rock backfill preparations. Tests using a bentonite additive in crushed granite and gneiss have shown it to be stable up to 100 °C. The swelling properties of the bentonite helps ensure that voids do not form in the rock backfill during emplacement [156]. At the Stripa mine in Sweden a buffer-backfill test used a mixture of sand and bentonite, which was shown to have suitable thermal and packing properties [157].

The backfill would probably be placed in layers, each undergoing physical compaction before the placing of more backfill material above. Where room is not available for physical compaction, work at' the Stripa mine in Sweden suggests that deposition spraying is an acceptable technique [157]. Quality control during backfill emplacement may in practice prove difficult to ensure.

Sealing of the sites may be done with concrete, bitumens, plastics, epoxy or silicate-based grouts, butyl rubber or corrugated non-ferrous metals such as copper or aluminium. These materials may also be used to separate backfill layers. While there is considerable knowledge as to how these various waterproofing materials perform in the short term there is not much information available as to how they may operate under longer-term repository conditions.

7 Seabed disposal of high-level radioactive waste

7.1 INTRODUCTION

As with on-land disposal seabed disposal of high-level radioactive waste is a multi-barrier concept. The concept envisages that waste would be vitrified and then packed into containers and overpacks before being placed on the seabed, or buried at varying depths within its soft sediment covering and the underlying lithified sediments and oceanic basalts. These various barriers, which are broadly similar to those relied upon in the land-based disposal concept, are further aided in the seabed disposal concept by the vastness of the oceans and their enormous dilution capacity. An appreciation of the seabed concept requires a knowledge of the nature of the seabed and its sediments, the development of appropriate emplacement technologies and an understanding of how escaping radionuclides will be transported and how they could affect the life of the oceans and beyond. About 70% of the globe is covered with seawater of varying depths, which might give the impression that the potential choice of suitable disposal sites would be great. This is not the case, however. The inland seas, continental margins and shelves are clearly unsuitable, being generally quite shallow and with important fisheries, oil and other mineral interests. It may also be considered that by definition such seas are unnecessarily close to land masses and man's habitation. Similarly it is unlikely that the seismically active mid-oceanic ridges would be suitable disposal sites. The search for seabed disposal sites is therefore confined mainly to the region between these two areas, that is in the area of the ocean basin floor composed of the deep ocean trenches and the surrounding abyssal hills and plains. In general, research has concentrated on the abyssal plains owing to their greater thickness of sediment, their relatively low economic importance, low biological activity and their seismic passivity. While emplacement on the seafloor is under consideration it is as well to record that the limitations on such a concept are more related to oceanographic considerations than to the lack of suitable geological conditions. While such a method has the advantage that canisters exposed at the seafloor will have the benefit of being cooled by the surrounding seawater and relatively low emplacement costs, once the canisters of waste are breached the waste form may be completely dissolved after a few years. The safety of the method will then depend upon an efficient dispersal and dilution of the dissolved waste by the ocean waters. According to the Institute of Oceanographic Sciences in 1979 [158] it was not possible at that time to say whether or not this could be achieved and at the time of writing this is still the case.

7.2 THE LONDON DUMPING CONVENTION

The disposal of nuclear waste beneath the high seas poses particular problems of a legal and organizational nature as a result of their 'common' ownership. Since 1949 the United Kingdom has dumped low-level and intermediate-level wastes into ten or so sites on the eastern side of the North Atlantic as shown in Fig. 7.1. Nuclear waste has also been dumped in these areas by a number of other countries including France, Belgium, Italy, Sweden, Switzerland, the Federal Republic of Germany and The Netherlands. The dumping of these wastes is now controlled by the provisions of the Convention on the Prevention of Marine Pollution by Dumping of Wastes and Other Matter [160], which is more commonly known as the 'London Dumping Convention'.

As Fig. 7.2 shows considerable quantities of low- and medium-level radioactive waste material have been placed in the North Atlantic since 1949 though the London Dumping Convention has only been in force since 1975. The waste is generally packed into steel drums, often in a concrete matrix, and simply dropped over the side of the dumping vessel. The main function of the steel drum is to ensure that the container reaches the ocean floor intact as well as making the waste easier to handle. The dumping of these relatively low-level, non-heat-generating wastes should not be confused with the dumping of high-level nuclear waste which is considered unsuitable for dumping in a similar manner by Annex I of the London Dumping Convention.

Article III of the London Dumping Convention specifically defines the term 'dumping' as 'any deliberate disposal at sea of wastes or other matter from vessels, aircraft, platforms, or other man-made structures at sea'. The legal question presented therefore is whether or not the actual burial of high-level nuclear wastes within the seabed would constitute 'dumping' as outlined in the convention. The general conclusion seems to be that the convention does not preclude disposal in the seabed provided it is carried out in an environmentally sound and safe manner [161].

It is also likely that the 1982 Convention on the Law of the Sea [162] requires care to be taken when disposing of waste on or under the seabed. It may also be argued that the general principles of

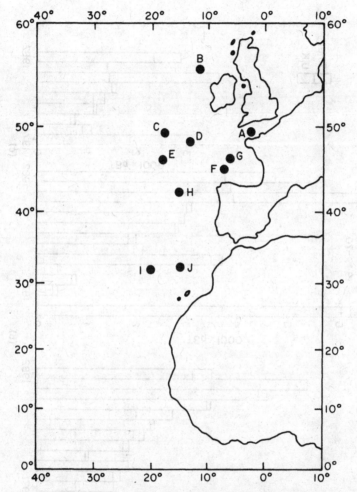

Figure 7.1 Dump sites used by the United Kingdom for the disposal of nuclear wastes [159]. Dates of use and total number of years employed (in brackets) are: A (Hurd Deep) 1950–1963 (14); B 1951, 1953 (2); C 1969 (1); D 1949, 1965–1966, 1968, 1970 (5); E 1971–1982 (12); F 1963–1964 (2); G 1962 (1); H 1967 (1); I 1957–1958, 1961 (3); J 1955 (1).

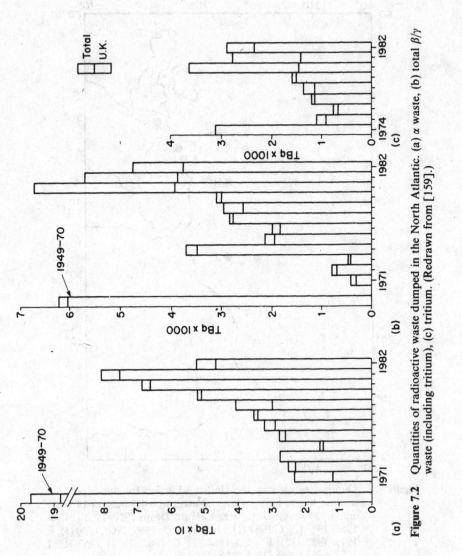

Figure 7.2 Quantities of radioactive waste dumped in the North Atlantic. (a) α waste, (b) total β/γ waste (including tritium), (c) tritium. (Redrawn from [159].)

international law impose an environmental standard of care or duty upon all nations using the world's oceans and perhaps a measure of responsibility and liability for any damage that may result from such use.

7.3 CRITERIA FOR THE SELECTION OF SEABED DISPOSAL SITES

The NEA [161] suggest that there are two main criteria for the selection of seabed disposal sites: stability and barrier. In other words, the geological formations should be stable and predictable while the rock or sediment medium should have characteristics which make it an effective barrier to the release of radionuclides.

Stability and predictability factors

Site area this will depend largely on the disposal method adopted. As Fig. 7.3 shows the area required for seabed disposal varies markedly according to the spacing between penetrators and the number of waste canisters per penetrator. To allow for inaccuracies in emplacement and for adequate heat dissipation a horizontal spacing

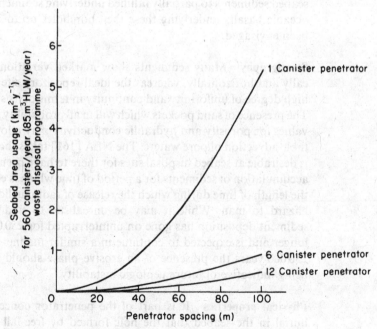

Figure 7.3 Seabed area usage for penetrator disposal [163].

between waste canisters of about 250 m is envisaged. The NEA are of the view that a minimum area of about 100 km^2 capable of taking 100 000 waste canisters should be considered [161].

Site bathymetry   the main emphasis here is on choosing sites free from turbidity currents and sediment slumping. In general these requirements are likely to be found where the seabed slopes very gently, ideally no more than 1:1000 [158, 164] in areas distant from steep canyons and slopes which could give rise to mass movement of sediment. Sites should also be examined for evidence of recent erosion, the presence of which should be taken as potentially excluding them from further consideration.

Sediment thickness and structure   On the assumption that the most shallow penetrators will bury themselves at least 30 m, and accepting that a comparable thickness of sediment should exist below the canister as well as above, then a minimum thickness of about 60 m would be required. For burial below the soft seabed sediment the depth of boreholes will to a large extent be determined by economic factors related to the costs of drilling and emplacement, but it is unlikely that sediment thicknesses of less than 400 m would be satisfactory if more than one canister is to be emplaced in each hole. If it is intended to drill right through the relatively unconsolidated seabed sediments to partially lithified underlying sediment or into the oceanic basalts underlying these then boreholes up to 3·5 km have been envisaged.

Stratigraphy   Many sediments show marked variation both vertically and horizontally, whereas the ideal repository site will show a high degree of uniformity and continuity in terms of its stratigraphy. The presence of sand pockets which will in all probability show higher values for porosity and hydraulic conductivity may allow unpredictable advection of pore waters. The NEA [161] takes the view that it is desirable at seabed disposal sites for there to have been continuous accumulation of sediments for a period of time which is equivalent to the length of time during which the release of radionuclides may be a hazard to man. While it may be unrealistic to find sites where sediment deposition has gone on uninterrupted for 250 000 years or longer and is expected to continue in a similar manner for another 250 000 years the presence of an erosive phase should be taken as being indicative of future geologic instability.

Physical properties   It is part of the penetrator concept of waste burial in the seabed that the hole formed by free-fall penetrators should be self-sealing. For this to take place effectively requires the sediments to be viscoelastic and plastic to a considerable degree. Clearly the presence of significant boulder or gravel layers will make the effective self-sealing of holes less likely, as well as potentially reducing the depth and accuracy of penetrator emplacement. It is also important to consider the effect of heat and radiation upon the geochemical character of the sediment to ensure that it continues to be an effective retention and barrier medium.

Barrier property factors

The sorption of clays As with the use of land-based clay repositories it is to be expected that the fine clay particles within the unconsolidated seabed sediments will be able to reduce the rate of migration of escaping radionuclides by the process of sorption. Clearly there would be advantages to the use of sites with a predominance of clay as opposed to sand or other materials with a lower sorption capacity. Sand and silt particles tend to have lower adsorption capacities owing to their reduced surface areas for a given volume of sediment and their chemically less active surfaces. In clay sediments precipitation and ion exchange can be expected to result in a net removal of ions from the pore water solutions. In deep ocean carbonate zones, however, the carbonate matrix is slowly dissolving, resulting in a net release of ions and thus a lower adsorption capacity for the solid phase. The reasons for this dissolution of carbonate material at depth are two-fold. Seawater at depth is very cold, probably only $2°C$ or so at the depths under consideration. Colder water contains more dissolved carbon dioxide than does the warmer surface water where the carbonate-secreting organisms originally lived. The increased carbon dioxide results in an increase in the rate of dissolution of the carbonate material at depth. Increased hydrostatic pressure at depth of between 400 and 600 atmospheres in the main study areas also leads to increased dissolution of carbonate material, which is unlikely to survive in any quantity below about 5000 m of sea water.

Porosity It is expected that the main migration of radionuclides away from buried canisters would be through the sediment pore waters. In addition radioactive material may be transported by mobile pore water. Pore water may be mobilized by compaction of unconsolidated sediment or by the release of thermal energy from the contained waste. The NEA have suggested as a guide that the rate of pore water advection at any particular site should be less than the rate of transport by other mechanisms such as diffusion. The chemistry of the pore waters will also be important. The solubilities of some ions is increased under low oxidation conditions while others show similar increased solubility under high oxidation levels. To give the maximum chance of retarding the escape of certain radionuclides through the entire thickness of overlying clay sediment it may be advantageous to use sites with a range of redox conditions from sediment level to sediment level.

Bioturbation   The presence of benthic burrowing organisms may result in unpredictable sediment characteristics at depth. While animals are unlikely to burrow from the surface to the likely depths of canister burial, successive burrowing at previous seabed sediment levels has to be taken into account. The NEA view is that the presence of burrows in excess of 1% of the total sediment volume to the depth of burial should be considered an undesirable characteristic.

Depth of sea   In general it is most likely that sites with a cover of at least 4000 m of seawater will be chosen.

7.4   SITES UNDER CONSIDERATION FOR SEABED DISPOSAL

Studies to date have concentrated on a number of widely dispersed sites in the North Atlantic and the North Pacific, the principal ones of which are shown in Fig. 7.4. The Seabed Working Group of the NEA has reviewed some 15 areas in the North Atlantic and five areas in the Pacific Ocean. As a result the following areas would appear to meet with the minimum criteria set out above:

Sites in the North Atlantic:
1. Great Meteor East in the Madeira Abyssal Plain
2. Southern Nares Abyssal Plain
3. King's Trough Flank
4. Cape Verde Rise–CV2

Sites in the Pacific:
1. B1
2. C1
3. E2

See Fig. 7.4 for the locations of these code identification names. Detailed descriptions of these sites, as far as they are known, have been put together by the NEA from a variety of sources and these are summarized in [161].

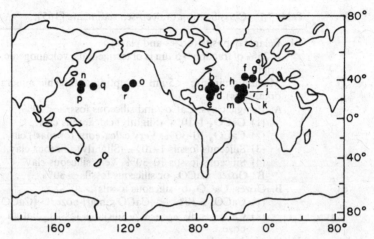

Figure 7.4 Seabed disposal sites under investigation. (Redrawn from [161].) Areas in the North Atlantic: (a) SBR: Southern Bermuda Rise; (b) NBR: Northern Bermuda Rise; (c) SSAP: Southern Sohm Abyssal Plain; (d) NNAP: Northern Nares Abyssal Plain; (e) SNAP: Southern Nares Abyssal Plain; (f) KTF: King's Trough Flank; (g) IB1: Iberia 1; (h) GMW: Great Meteor West; (i) GME: Great Meteor East; (j) Madcap: SW Madeira Abyssal Plain; (k) CV1: Cape Verde Abyssal Plain (East); (l) CV3: Cape Verde Abyssal Plain (West); (m) CV2: Cape Verde Rise. Areas in the North Pacific: (n) B1: PAC I (1100 km east of Japan); (o) MPG II: PAC II (1900 km northeast of Hawaii); (p) C1: PAC I (1100 km southeast of Japan); (q) E2: Pac I (2000 km southeast of Japan); (r) MPG I: PAC II (1100 km north of Hawaii).

7.5 THE NATURE OF THE SEABED SEDIMENTS

The pelagic sediments found in the deep oceans far removed from the continents and their sediment-laden rivers consist mainly of fine clay particles with additional amounts of biologically produced carbonate and silicate material in biologically productive areas. Mixed in with these clays and oozes may be sand horizons, the result perhaps of turbidities, and layers of authigenic minerals resulting from post-deposition chemical change and early lithification processes. For the most part pelagic deposits have a mineralogy derived from volcanic eruptions or have a secondary origin such as the red clays which are formed *in situ*. The remains of pelagic organisms that have lived in surface waters before their death and fall to the bottom of the oceans also occur in large quantities in mid-oceanic deposits. Pteropod, Diatom, Globigerina and Radiolarian oozes are all found in the deep oceans. Table 7.1 gives a classification of deep-sea pelagic sediments, while Fig. 7.5 shows the global distribution of these sediments. It is

Table 7.1 Classification of deep-sea sediments [165].

1. Pelagic deposits (oozes and clays)
   $<25\%$ of fraction >5 μm is of terrigenous, volcanogenic, and/or neritic origin.
   Median grain size <5 μm (excepting authigenic minerals and pelagic organisms).
   A. Pelagic clays. $CaCO_3$ and siliceous fossils $<30\%$.
      (1) $CaCO_3$ 1–10%. (Slightly) calcareous clay.
      (2) $CaCO_3$ 10–30%. Very calcareous (or marl) clay.
      (3) Siliceous fossils 1–10%. (Slightly) siliceous clay.
      (4) Siliceous fossils 10–30%. Very siliceous clay.
      B. Oozes. $CaCO_3$ or siliceous fossils $>30\%$.
   B. Oozes. $CaCO_3$ or siliceous fossils $>30\%$.
      (1) $CaCO_3 > 30\%$. $<\frac{2}{3}CaCO_3$: marl ooze. $>\frac{2}{3}CaCO_3$: chalk ooze.
      (2) $CaCO_3 < 30\%$. $>30\%$ siliceous fossils: diatom or radiolarian ooze.
II. Hemipelagic deposits (muds)
   $>25\%$ of fraction >5 μm is of terrigenous, volcanogenic, and/or neritic origin.
   Median grain size >5 μm (excepting authigenic minerals and pelagic organisms).
   A. Calcareous muds. $CaCO_3 > 30\%$.
      (1) $<\frac{2}{3}$ $CaCO_3$: marl mud. $>\frac{2}{3}$ $CaCO_3$: chalk mud.
      (2) Skeletal $CaCO_3 > 30\%$: foram $\sim$, nanno $\sim$, coquina $\sim$.
   B. Terrigenous muds. $CaCO_3 < 30\%$. Quartz, feldspar, mica dominant. Prefixes: quartzose, arkosic, micaceous.
   C. Volcanogenic muds. $CaCO_3 < 30\%$. Ash, palagonite, etc., dominant.
III. Pelagic and/or hemipelagic deposits.
   (1) Dolomite–sapropelite cycles.
   (2) Black (carbonaceous) clay and mud: sapropelites.
   (3) Silicified claystones and mudstones: chert.
   (4) Limestone.

also of note that ice-rafted debris is to be found among these pelagic sediments. In this respect it should be borne in mind that during the Pleistocene ice-rafted material would have been deposited much nearer to the equator than is the case today [167]. The rate of accumulation of pelatic sediments is very slow relative to areas nearer the continental land masses and is of the order of only a few mm per thousand years. Reading [166] quotes sedimentation rates of about 5×10^{-5} to 5×10^{-4} cm yr^{-1}, while Berger [165] gives Table 7.2 covering accumulation rates for various pelagic facies. Until recently it was assumed that the pelagic sedimentary column was complete, but this is apparently not the case. Results obtained by the Deep Sea Drilling Project have shown that in many ocean sedimentary columns there are unconformities [168, 169] of varying time gaps.

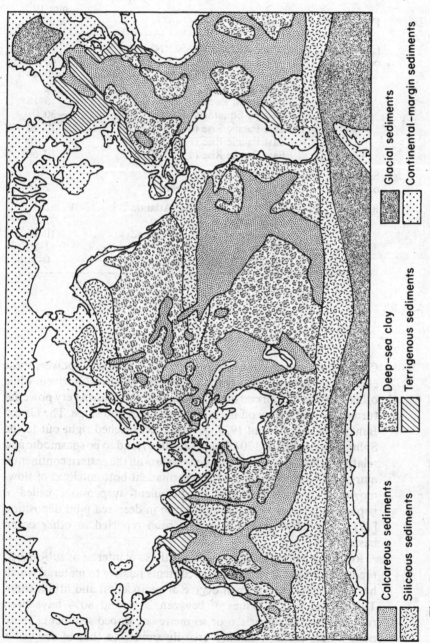

Figure 7.5 Global distribution of principal types of pelagic sediment on the ocean floors. (Redrawn from [166]).

Calcareous sediments	Deep-sea clay
Siliceous sediments	Terrigenous sediments
Glacial sediments	
Continental–margin sediments	

Table 7.2 Rates of accumulation of Recent and sub-Recent pelagic facies [165].

Facies	Area	mm/10^3 years
Calcareous ooze	North Atlantic (40–50 °N)	35–60
	North Atlantic (5–20 °N)	40–14
	Equatorial Atlantic	20–40
	Caribbean	~28
	Equatorial Pacific	5–18
	Eastern Equatorial Pacific	~30
	East Pacific Rise (0–20 °S)	20–40
	East Pacific Rise (~30 °S)	3–10
	East Pacific Rise (40–50 °S)	10–60
Siliceous ooze	Equatorial Pacific	2–5
	Antarctic (Indian Ocean)	2–10
Red clay	North and Equatorial Atlantic	2–7
	South Atlantic	2–3
	Northern North Pacific (muddy)	10–15
	Central North Pacific	1–2
	Tropical North Pacific	0–1

Clearly there are abyssal currents at work. Quite what powers these currents is not entirely clear. Some may be density-driven while others may be the tail end of the reasonably well known very powerful turbidities which flow off the continental shelf and slope. The Grand Banks turbidity flow of 1929, for instance, reached right out to the Sohm Abyssal Plain [170]. Such turbidities tend to be spasmodic and relatively short-lived but in the deep waters off the eastern continental margin of North America a semi-permanent bottom cloud of slow-moving clay-sized sediment in turbulent suspension, called a nephaloid layer, may be of importance in deep-sea mud deposition [171]. Nephaloid layers have also been reported in other ocean basins [172].

Relatively young sediments such as those of interest as sub-seabed repositories have very high water contents relative to material which has undergone even a small degree of compaction and lithification. Typical wet-weight values of between 20% and 80% have been recorded for the upper ten or so metres of seabed sediment. These pore waters retain a salinity broadly similar to that of sea water though they may be enriched with dissolved ions as a result of redox-controlled diagenetic processes. The strength of sediments is likely to vary from site to site to a marked degree but a range from 1 to 2·4 times the depth of burial expressed as kN m^{-2} is probably reasonable. Table 7.3 shows the main sediment and other characteristics of the 15 sites investigated by the Seabed Working Group of the NEA, while Table 7.4 shows sediment and pore water analyses for two cores from the Sohm Abyssal Plain. Figure 7.6 gives porosity and permeability data for some North Atlantic sediments. Despite relatively high porosity values the permeability of deep-sea sediments tends to be very low.

7.6 EMPLACEMENT TECHNIQUES

Up to 19 different canister emplacement techniques have been considered [174]. In general these can be subdivided according to whether the hole formation and the emplacement of the canister are combined in one operation or not. Table 7.5 summarizes the various emplacement techniques under investigation. Manschot, Callanfels and Van [175] have described the penetrator emplacement concept which at its simplest uses free-fall penetrators which come to rest at a depth in the seabed dependent upon the penetrator design and the nature of the seabed. The drilled emplacement method, by comparison, envisages the drilling of a hole into the sediment or into the

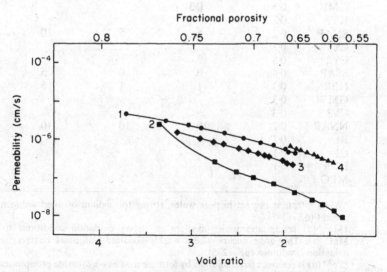

Figure 7.6 Permeability versus void ratio of some North Atlantic sediments. 1 Marl 52% CaCO$_3$ horizontal flow. 2 Red pelagic clay vertical flow. 3 Marl 50% CaCO$_3$ vertical flow. 4 Marl 64% CaCO$_3$ vertical flow. (Redrawn from [173].)

Table 7.3 Sediment components [161].

Area name	Water content (% wet)	Clay mineral content (% dry)	Sand–silt content (% dry)	Carbonate content (% dry as $CaCO_3$)
GME	40–60	10–30	10–30	50–70
KTF	40–60	10–30	10–30	60–80
SNAP	40–60	90–100	0–5	0–5
CV1	50–70	20–40	10–30	40–60
CV2	40–60	10–30	0–20	60–80
SSAP	35–55	70–90	0–20	0–20
NBR	35–55	40–60	10–30	10–30
GMW	40–60	0–10	0–10	80–100
CV3	40–60	20–40	20–40	40–60
NNAP	35–55	90–100	0–5	0–5
Bl	50–70	90–100	0–5	0–5
MPG II	50–70	90–100	0–5	0–5
Cl	50–70	90–100	0–5	0–5
E2	50–70	90–100	0–5	0–5
MPG I	50–70	90–100	0–5	0–5

Area name	Organic carbon (% dry)	Thickness of nitrate zone* (m)	Depth to MnO_2 zone† (m)	Depth to Fe redox discon.‡ (m)
GME	0·3	0·3		
KTF	0·2			
SNAP	0·2	5	5	10
CV1	0·3			
CV2	0·5	2		
SSAP	0·4	0	0	5
NBR	0·3	1	1	5
GMW	0·3			
CV3	0·3			
NNAP	0·2	10	10	10
B1	0·2			
C1	0·1			
E2	0·1			
MPG I			15	24

*When nitrate is present in pore water, strong to medium oxidized sediments are present ($pE = 5$–7).

†Mn^{2+}(d) precipitates under medium to strong oxidation conditions to form MnO_2(s). This zone occurs where weakly oxidized sediments contact medium oxidation conditions ($pE = 2$–5).

‡Fe^{2+}(d) is removed from solution by forming iron oxy-hydroxide precipitates. This generally occurs where reduced sediments contact weekly oxidized sediments ($pE = -3$ to $+2$).

See Fig. 7.4 for locations.

Table 7.4 Sediment and pore water analyses for two cores from the Sohm Abyssal Plain (summary of data for 48 samples from depths of 0 to 12 m in the sedimentary column).

Parameter	Concentration unit	Sample type	Mean	Standard deviation
Depth	cm	Core	438	430
Clay	%	Dry sediment	82	26
Organic carbon	%	Dry sediment	0·43	0·15
Water content	%	Wet sediment	42	12
$CaCO_3$	%	Dry sediment	11	4
Total Fe	%	Dry sediment	4·5	1·0
Total Mn	p.p.m.	Dry sediment	671	94
Total Zn	p.p.m.	Dry sediment	95	23
Total Cu	p.p.m.	Dry sediment	23	7
Total Cr	p.p.m.	Dry sediment	89	23
Total Ca	%	Dry sediment	4·2	1·8
Total Si	%	Dry sediment	25·3	3·6
Total Al	%	Dry sediment	8·1	1·3
Total Ni	p.p.m.	Dry sediment	60	20
pH2 leachable Fe	p.p.m.	Dry sediment	1470	510
pH2 leachable Mn	p.p.m.	Dry sediment	232	63
pH2 leachable Zn	p.p.m.	Dry sediment	5·9	2·7
pH2 leachable Cu	p.p.m.	Dry sediment	4·6	2·5
pH2 leachable Cr	p.p.m.	Dry sediment	4·4	1·5
Reducing leachable Fe	p.p.m.	Dry sediment	1660	660
Reducing leachable Mn	p.p.m.	Dry sediment	262	79
Reducing leachable Zn	p.p.m.	Dry sediment	6·8	3·3
Reducing leachable Cu	p.p.m.	Dry sediment	4·8	3·4
Reducing leachable Cr	p.p.m.	Dry sediment	7·3	2·7
E_H	mV	Wet sediment	191	77
pH		Wet sediment	7·8	0·2
Alkalinity	meq $^{-1}$	Pore water	5·5	1·5
Fe	μM	Pore water	24·9	29·4
Mn	μM	Pore water	59·2	23·5
Silicate	μM	Pore water	228	54
Phosphate	μM	Pore water	8·9	4·8
Nitrate	μM	Pore water	9·7	2·2
Zn	nM	Pore water	300	90
Cu	nM	Pore water	30	3
Cr	nM	Pore water	0·5	0·1
Na	mM	Pore water	465	20
K	mM	Pore water	10·2	0·5
Ca	mM	Pore water	10·8	0·5
Mg	mM	Pore water	50·6	2·2
Al	nM	Pore water	40	5

Table 7.5 Emplacement techniques [161].

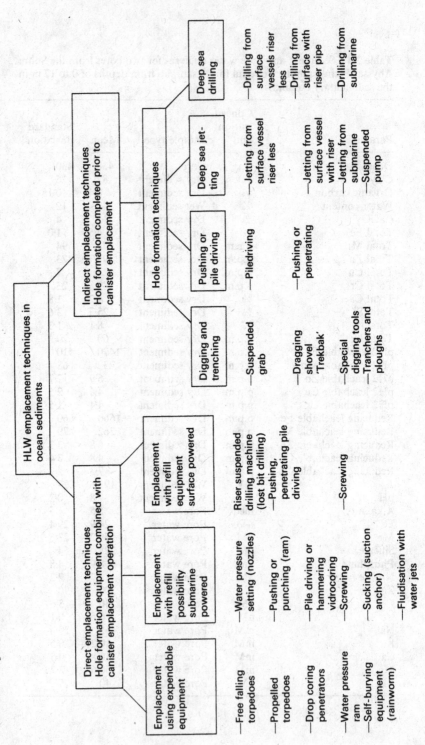

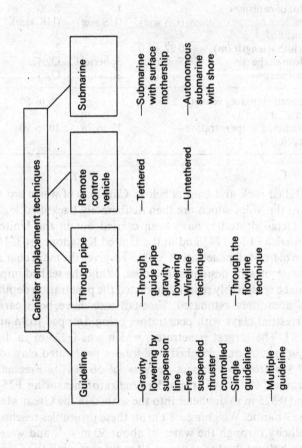

Table 7.6   Penetrator sizes [161].

Penetrator details	US [177]	US [176]	US [163]	
Overall length (m)	5·0	1·5–10·5	4·3–30·0	
Outside diameter (m)	0·4	0·3–0·7	0·43–0·98	
Total weight (tonne)	3·44	0·5–16·4	1·8–153·2	
No. of canisters	1 (US size)	1 (US size)	2–20 (UK size)	
Fins no.	4	4	3	
Width × length (m)	0·22 × 1·0	?	?	
Nose shape	Spherical	Spherical	Ogival	
Tail shape	Square	Square	Ogival	
Seabed impact velocity (m s⁻¹)	—		25 to 50	21 to 77
Predicted penetration— depth (m)		12 to 78	10 to 90	

lithified rock and basalts below. Canisters of waste are then guided into the holes which are then deliberately backfilled.

Detailed studies have been carried out in the United States of America [176, 177] and in the United Kingdom [163, 164] into the use of free-fall penetrators. Table 7.6 gives the main characteristics of the penetrator designs considered. While the seabed impact velocity can be reasonably reliably predicted the penetration depths cannot be so accurately estimated. Free-fall tests have been carried out on terrestrial clays with penetrators being dropped from aircraft [178, 179]. The largest penetrator, which was 0·225 m in diameter and 2·84 m long and weighed 0·68 tonnes penetrated clay to a depth of 23·3 m from an impact velocity of 56 m s⁻¹. Freeman, Murray, Francis [180] have dropped a penetrator measuring 3·25 m in length and 0·325 m in diameter into the seabed in the Great Meteor area of the Atlantic. Weighing 1·8 t in air these projectiles reached a terminal velocity through the water of about 50 m s⁻¹, and were tracked to sub-bottom depths of about 30 m by recording the Doppler-shifted frequency of a continuous 12 kHz acoustic source embedded in the tail.

Penetration results on other sediments have been recorded using instrumented seabed penetrators in relatively shallow waters. Penetration depths of up to 12 m have been achieved in calcareous ooze using an instrument with a diameter of 0·9 m, a length of 2·4 m and a weight of about 0·14 tonnes. The terminal velocity was about 26 m s⁻¹ [181]. A boosted instrumented seabed penetrator capable of impact velocities of 100 m s⁻¹ achieved an embedment depth of

37 m from an impact velocity of 88 m s^{-1}. These results suggest that the penetrator concept is a feasible one as far as self-burial requirements are concerned.

Penetration depths can also be predicted from empirical equations derived from land-based experimentation. While it will be necessary to supplement and refine these equations with seabed-derived calibration data there seems every prospect that reasonably reliable estimates of penetrator burial will be possible. As an example of the degree of control which may be possible, based in this case on the assumed design parameters given in Table 7.7, Fig. 7.7 shows how burial depths may vary according to the length, weight and cross-sectional area of penetrators. The reliability of such predictions in practice will to a large extent depend upon an accurate assessment of the sediment characteristics, requiring the development of new *in situ* deep-sea instruments to measure shear strength. Good borehole records will also be needed; research suggests, for instance, that sand layers and boulders will decrease penetration depths [163, 176].

Hole closure behind the penetrator will depend upon the dynamic pressure reduction or suction behind the penetrator. This is proportional to the velocity of penetration and needs to be as high as is reasonably practical. The amount of pressure required to fill the hole behind the penetrator is proportional to the undrained shear strength of the sediment. The magnitude of the deceleration of the penetrator will also influence the hole closure as this will result in a pressure increase. In general, preliminary calculations and numerical modelling suggest that hole closure should be achieved. While full-scale trials will be required in any proposed seabed disposal area initial tests in the field and using kaolin in the laboratory suggest that even at very low velocities hole closure can be achieved [182].

Table 7.7 Design parameters used for embedment sensitivity analysis [161].

Design parameter	US [176]	UK [163]
Soil strength (c_u) at sub-seabed depth Z(m) (kN m^{-2})	$c_u = 19.54 \times \log_e (1 + 0.1189Z)$	$c_u = 1.5Z$
c_u at $Z = 60$ m (kN m^{-2})	41	90
Soil sensitivity	?	5
Dynamic end bearing coefficient (N_{ed})	9	18
Allowance for hole closure and rate effects	No	Yes
Dynamic adhesion factor (α_d)	0.26	0.4
Inertia drag in soil	Not included	Not included

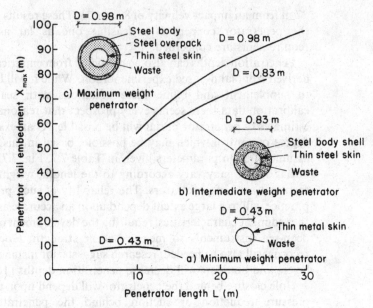

Figure 7.7 Relationship between penetrator design parameters and depth of burial [161].

The drilled emplacement method will to a large extent depend upon existing drilling technology. Ships such as *Glomar Challenger* already have considerable experience of drilling in water up to 6000 m deep and have developed holes up to 1750 m below the seabed. No riser is needed as water is used for the drilling fluid and hole debris is left on the seabed. The NEA are of the view that it should be possible to drill uncased holes up to 0·5 m in diameter in the lithified rock underlying the softer surface sediments, while cased holes up to 0·9 m could be developed in the softer surface sediments [161]. The method of transferring the waste flasks to the seabed hole is currently under investigation, but one promising idea is to place the waste canisters in 15 m long pipes, with each pipe taking about ten canisters. These pipes are then connected together to form a long pipe string which is then lowered on a wire or a conventional pipe string to a relocation cone and then into the drillhole. The necessary relocation techniques using sonar and TV cameras already exist to achieve this satisfactorily.

Unlike the penetrator case, backfill has to be deliberately put into position in the drilled emplacement concept. Four major backfill locations have been identified by St. John, Mayne and Hills [183].

These are:

'–the space or annulus between the waste canister and the borehole wall, assuming both container and borehole are circular in section; the average annulus width will be small (c 0·1 m) compared with its length which is that of the container (c 1·5 m), and will vary if the container is not concentric in the borehole,
–the vertical space in the borehole between the containers (from 0 m upwards); this region may be empty of solids before backfilling or may contain material connecting the containers,
–the vertical space in the borehole between the highest container and the seabed (c 500 m),
–the volume of sediment in the disturbed zone around the borehole, the condition and properties of which may differ from the bulk sediment due to the construction and presence of the borehole; the nature of the disturbance may be cracking, overbreaking or dilation of the sediment which may cause local changes in its hydraulic conductivity, water content, and permeability.'

The functions of the backfill material are to locate the canisters within the borehole, to control the vertical flow of water within it, to conduct heat away from the waste canisters and to limit the migration of radionuclides. Broadly these requirements are similar to those already discussed for on-land disposal.

An ideal backfill material will display strength, elasticity, a high sorption capacity, good thermal conductivity, low chemical reactivity, dimensional stability, durability and low permeability and porosity. It should also be capable of easy emplacement and rigorous quality control. The most likely backfill materials to be used include hydraulic cements, bentonite clays and processed host rock. Of these, cement-based materials may be most suitable as they are known to be long-lasting if properly emplaced, they can cope with hot-rock conditions, they are readily available and their setting time, pumpability and chemical compatibility can all be changed to suit particular hole conditions [183].

7.7 THE WASTE FORM AND THE CANISTER

In a similar manner to the land-based waste disposal concept it is envisaged that waste will be incorporated into a borosilicate glass. A few experiments have been carried out on the susceptibility of these glasses to leaching with seawater as opposed to fresher or distilled waters [184, 185]. Using both high flow rates and large volumes of water relative to the glass surface areas suggests that there are no marked differences between the leach rates for seawater and groundwaters. Using a similar approach and assuming the worst possible scenario, that of a canister failing to penetrate the seabed sediment and shattering into 10 cm diameter spheres on cooling, the NEA estimate the leaching loss would amount to about 0·1 g per day [161].

As with the land-based concept of disposal the projected time period for containment within the waste canister can vary. It may be appropriate in certain situations to rely upon the nature of the waste form itself and the surrounding sediment and seawater cover to provide the necessary degree of isolation. Where it is thought necessary to control the migration of cesium-137 and strontium-90 during the period when the waste is at a significantly higher temperature than the surrounding sediment then a canister containment period of 500 years or longer may be necessary. Two canister concepts are under investigation to provide for the longer periods of containment required. The first uses a relatively thin but very corrosion-resistant metal such as a titanium-based alloy, while the second works on a sacrificial principle, utilizing a thick overpack of cheaper metal such as mild steel, and accepts that corrosion will take place at fairly high rates, hence the initial much greater container thickness.

Using an alloy known as Ticode prepared by Timet Industries and assuming a canister–sediment interface temperature of between 200 and 250°C workers in the United States [186–188] have shown corrosion rates of about 1 μm yr^{-1} are to be expected. γ radiation may increase these corrosion rates by a factor of two but stress corrosion does not appear to be a factor with Ticode containers. Tests have also suggested that H_2 embrittlement which can pose problems in titanium alloys is unlikely to pose problems where pH conditions below 0·5 occur. Corrosion rates for carbon steel are estimated at about 23 μm yr^{-1} for general corrosion. This when considered along with additional pitting and crevice corrosion gives a maximum corrosion reduction in the overpack of 65 mm over a 500-year period. A container thickness of about 75 mm would therefore appear suitable in the circumstances. Rather than carbon steel, which is susceptible to hydrogen embrittlement, it may be preferable to use soft iron with a similar 75 mm allowance for corrosion.

Canisters must also remain intact under extreme hydrostatic pressures. These are likely to be around 60 N mm^{-2} at the sea depths under investigation in the abyssal regions of the oceans. The hydrostatic resistance requirement can be met either by the use of a rigid container or by the use of one which would deform under hydrostatic pressure, transferring the load in part to the container contents. Pressure equalization is not considered as a feasible option as it would of necessity involve an opening between the inside of the waste canister and the outside environment. Possible pressure-resistant penetrator designs are given in Fig. 7.8.

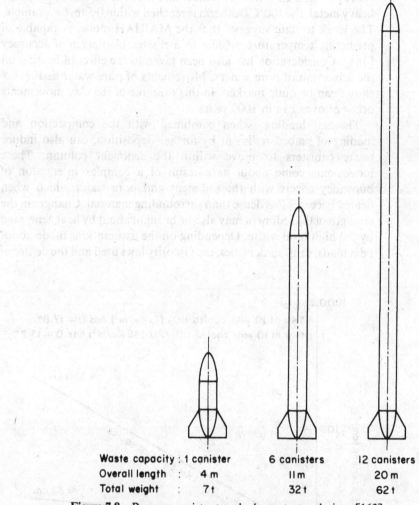

Waste capacity :	1 canister	6 canisters	12 canisters
Overall length :	4 m	11m	20 m
Total weight :	7t	32t	62t

Figure 7.8 Pressure-resistant seabed penetrator designs [163].

7.8 THERMAL EFFECTS ON THE SEDIMENT

The oceans can be considered as an infinite heat sink. Heat reaching the seabed, which is at an almost constant 2 °C at the water depths envisaged for seabed disposal, should not present any problems. While heat can be transported through sediment in a number of ways work on two thermal transport models (COYOTE [190] and MARIAH [191]) shows that thermal conductivity is by far the most important parameter in the transfer of heat through seabed sediment. Thermal environment predictions have been made for various waste forms. The maximum sediment–waste interface temperature is likely to be about 250°C but this falls off rapidly away from the waste canister. For a 1·5 kW heat source, that is about 1·5 metric tonnes of heavy metal, the 100°C isotherm is reached within 0·5 m, for example. The work to date suggests that the MARIAH model is capable of predicting temperature profiles to a reasonable degree of accuracy [161]. Consideration has also been given to the effect of heating on the advection of pore waters. Movements of pore water as Fig. 7.9 shows can be quite marked; in this example of red clay movements occur of over 1 m in 1000 years.

Thermal loading, when combined with the compaction and loading of seabed sediment by further deposition, can also induce waste canisters to move within the sediment column. These movements come about as a result of a complex interaction of buoyancy effects with the sediment and pore water which when heated becomes less dense than surrounding material. Changes in the strength of the sediment may also be brought about by heat generated by the high-level waste. Depending on the assumptions made about heat loads, canister densities, the viscosity laws used and the degree of

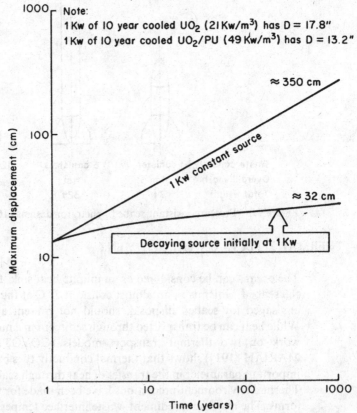

Figure 7.9 Pore water advection over time [161].

sophistication in modelling the separate sediment and pore water contributions, canister movement can be shown to range from as little as 1 cm over a 60-year period to as much as 36 cm over the same time [161, 192]. If one assumes that initial burial depths of between 10 and 30 m have been achieved these rates of movement should not pose difficulties.

7.9 THE SEDIMENT AS A BARRIER TO RADIONUCLIDE MIGRATION

It has already been pointed out from theoretical and experimental studies that clay minerals can be shown to have a considerable capacity to adsorb ions from solution. However, according to the NEA [161] 'the effectiveness for deep sea clays to act as a final barrier to radionuclides migrating from waste containers remains to be firmly established.'

Assessments of the ability of sediments to retard and retain radionuclides have been based mainly on the measurement of distribution and diffusion coefficients [193–195]. It is believed that radionuclide transport would take place almost entirely by molecular diffusion through the interstitial solution or pore waters and the initial work is encouraging.

For instance with a burial depth of 2 m, only the following isotopes breach the sediment barrier in significant amounts: ^{129}I, ^{237}Np, ^{239}Np, ^{240}Pu and ^{170}Pb. If the burial depth is increased to 10 m only ^{237}Np and ^{129}I remain a problem. Encouraging as results like these are, our understanding of how radionuclides actually move through, and react with, saturated deep-sea sediments is very limited and much more work remains to be carried out on the seabed sediments under as near natural conditions as possible. For instance, while the experimental work to date suggests that the fission products and actinides would decay to innocuous levels before significant migration has occurred, it is likely that significant quantities of certain anionic species such as I and Tc would escape through the sediment–water interface 5000–10000 years after release from the waste package. Clearly the nature of such radionuclide movements in reality needs to be properly understood.

7.10 THE TRANSPORTATION OF RADIONUCLIDES FROM THE SEABED–WATER INTERFACE TO THE FOOD CHAIN

Any radioactivity which reaches the seafloor from the buried canisters will disperse both horizontally and vertically, moved by ocean currents and by animals and plants which may ingest or pick up contaminants at one place and transport them to another. Minor movement may also be associated with sediment–water interchanges which may involve living material. A number of models of how waters circulate in the oceans and how food chains eventually reach man have been derived to predict the likely impact on man of an under-seabed disposal programme [197–201]. In 1976 the National Radiological Protection Board of the United Kingdom considered the consequences of dumping all the waste arising in the United Kingdom to the year 2000 beneath the seabed [202]. That is about 33·3 EBq of actinides and 3700 EBq of fission products. The worst analysis suggested that the public would be subjected to only 1% of the safe limits recommended by the ICRP. If the assumption was made that no deep-sea fish or plankton was consumed then the level of exposure reduced to only 0·04% of the ICRP dose limit. The recent Holliday Report [159] which considered the impact of past dumping of low- and intermediate-level radioactive waste in the North Atlantic would appear to confirm that radioactive material does not readily find its way back to man from the deep oceans. Table 7.8 shows the predicted peak annual doses from past dumping of radioactive waste in the North Atlantic. All this apart, it has to be said that our knowledge of deep ocean currents and how they change with altering climate, and their relationship with shallower surface waters where fish taken by man are more prevalent, is very limited. The traditional model to account for the circulation of the world's oceans is under review. No longer is it sufficient simply to account for oceanic circulation as a function of cold water forming at the poles and sinking to fill the deep ocean basins. When reviewing potential repository sites it will be necessary to have detailed knowledge of the bottom boundary layer and its movements as well as the incidence of more spasmodic erosive currents.

Equally our knowledge of deep-sea biota is very fragmentary. Despite the fact that there is an apparent exponential decrease in standing crop of biomass with depth which should limit the potential uptake of escaping radionuclides at depth, it is known that certain deep-sea species migrate vertically. These may link deep-water and surface-water food chains, albeit through a number of intermediate steps. The most commonly quoted example is the specimen of the amphipod *Eurythenes gryllus* taken at a depth of 800 m over a sounding in excess of 5000 m in the Pacific [203]. There are a number of reports in the literature of benthopelagic fishes with mesopelagic prey in their stomach contents, which have been interpreted as evidence for the occurrence of migrations by bottom-living fishes up into midwater to feed [204]. It is also necessary to take account of diurnal vertical migrations, that is the phenomenon found in many oceans whereby many species move from deep daytime depths up into the surface waters at night. Ontogenetic changes in living style and

Table 7.8 Predicted peak annual individual doses from past dumping [159].

Radionuclides	Annual individual dose* (μSv)	Exposure pathway†	Time of peak dose (years after dumping started)‡
'Actual pathways			
Americium-241	0·4	MOLL	85
Plutonium-239	0·01	MOLL, WEED	90
Cobalt-60	0·01	MINE	36
Plutonium-240	0·008	MOLL, WEED	85
Polonium-210	0·006	MOLL, CRUST	95
Plutonium-241	0·005§	BEACH	46
Plutonium-238	0·002	MOLL, WEED	67
Cesium-137	0·001	BEACH	60
Sum over all radionuclides and all actual pathways	0·8		
'Hypothetical' pathways			
Plutonium-239	0·008		36
Plutonium-241	0·006		36
Plutonium-240	0·005	FISH-D	33
Americium-241	0·003		41
Cesium-137	0·002		36

*Sum of the external dose equivalent to the whole body of the committed effective dose equivalent from intakes of radionuclides (but see §). This may be compared with the ICRP dose limit of 5 mSv and the average individual dose due to natural background radiation which is about 2 mSv.

†Pathways and symbols are as follows: consumption of deep fish (FISH-D); consumption of molluscs, crustacea and seaweed (MOLL, CRUST, WEED); external irradiation from working on the beach (BEACH); deep-sea mining of manganese nodules (MINE).

‡Dumping in the Northeast Atlantic started in 1949.

§Dose to skin from β radiation.

seasonal migrations also need to be assessed. While not necessarily present in large numbers, as Fig. 7.10 shows, some 95 species of benthopelagic fish occupy the entire column of water from the likely seabed depths of under-sea repositories all the way to the surface. Similar ranges can be demonstrated for the Asteroidea (starfish) and the Holothuroidea (sea cucumbers), amongst others. The links between all these species and the food chains leading to man are not at all clear. It seems, therefore, that while the deep seabed disposal option looks promising in many of its aspects much further research of field conditions is still required to be certain that the option is a safe one.

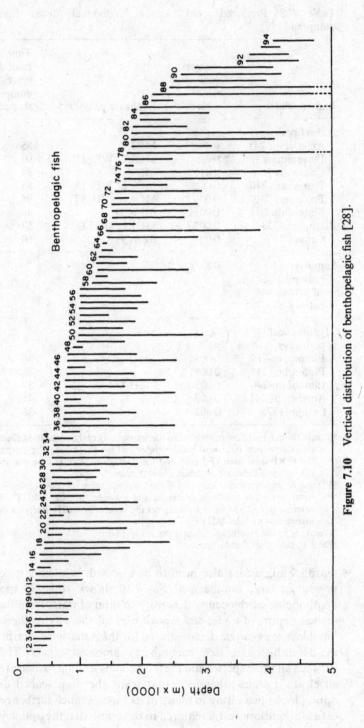

Figure 7.10 Vertical distribution of benthopelagic fish [28].

8 *Groundwater and its movement*

8.1 INTRODUCTION

According to Davis [206], 'although the escape of gas, erosion of the land surface by water, uptake by plant roots, excavation by explosions, and penetration by animals and man can all cause the undesired spread of radioactive waste, the most likely agent for this spread is commonly assumed to be groundwater that has passed through the waste'. This is a view widely repeated in the literature and is one of the most important factors to be taken into account when considering the effectiveness of the multi-barrier concept of geological disposal. A knowledge of the nature of groundwater, its occurrence and movement is therefore essential in determining the acceptability or otherwise of any particular HLW repository. While reference has been made to hydrogeological aspects within the context of earlier chapters it remains to provide additional data of a more general nature.

8.2 THE NATURE OF GROUNDWATER AND ITS ABILITY TO DISSOLVE GEOCHEMICAL MATERIAL AND RADIONUCLIDES

While water can occur in solid, liquid and gaseous phases it is as a liquid that it is most important in rocks, where as a very powerful solvent it has the ability to pick up and transport geochemical material where rock conditions permit. The main physical properties of water are given in Fig. 8.1, from which it can be seen that many of these are temperature dependent. Water temperature will to a large extent control the solvent action of groundwater, while viscosity will control its mobility in part.

Pure water dissociates to a small extent into H^+ and OH^- ions, with the degree of this dissociation being known as the pH of the water. The pH is in fact a measure of the hydrogen-ion concentration expressed in terms of the logarithmic value of this concentration. The pH of pure water at 24°C is 7.00. At 0°C this rises to 7.47 while at 60°C it falls to 6.51. The solubility of many species will alter with changing pH of the groundwater. Solubility will also be influenced by the extent of the contact surface over which dissolution is able to take place, the

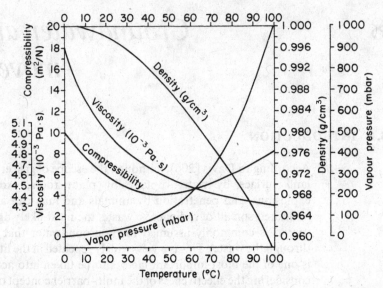

Figure 8.1   Temperature-dependent water characteristics (density, vapour pressure, viscosity and isothermal compressibility) [25].

time of contact, the difference between the highest possible solution concentration and the initial concentration in the solution, and whether the solute is constantly being replaced or whether transport of dissolved material is only able to take place by means of diffusion. Solubility will also be influenced by ionic activity and interactions between different substances both in the rock subject to dissolution and in the solute. For a detailed discussion of solutions and groundwater see Matthess [25].

The solubility of many elements is also dependent upon the oxidation state, which is controlled by the redox potential (E_h) and the pH of the environment. The redox potential is a measure of the relative state of the oxidation or reduction in an aqueous system. A number of workers have examined the use of E_h–pH diagrams to assess the solubility and hence the retention and migration of certain radionuclides [10, 207]. While such diagrams can produce realistic results when comparing predicted results for retention and migration of radionuclides with measured results in the field [208] to be reliable they need precise data relating to ionic strength, temperature, pressure and chemistry. More often than not such data are not present at a sufficient level of accuracy, in which event while E_h–pH diagrams can still be usefully invoked to gain an understanding of how various systems work, great care needs to be taken when drawing firm conclusions from them.

The solubilities of various components of HLW have already been given elsewhere together with analyses of a number of different groundwaters. Nevertheless the importance of groundwater composition and temperature on its solvent ability needs to be underlined and this is well illustrated by the recent work of Rees, Cleveland and Nash [209]. They conducted leaching experiments on radioactive waste glass using basalt, granite, shale and tuff groundwaters, amongst others, at 25 and 90°C, the analyses of which are given in Table 8.1. The amount of plutonium leached from the vitrified waste was measured after various time periods up to 360 days. Rees, Cleveland and Nash found that the quantity of waste leached depended on the chemical composition of the groundwater used as the leachant. The basalt water, for example, was the most effective leachant for plutonium at 90°C. The tuff water, while being very effective at dissolving the glass matrix, was not very effective at removing plutonium from the glass. The water from the shale leached the least material from the glass. As Table 8.2 shows the initial composition of the groundwaters also undergoes quite marked changes as a result of leaching the vitrified waste, and this is clearly another factor which needs to be allowed for when considering groundwater–waste interactions.

8.3 MOVEMENT OF GROUNDWATER

Groundwater possesses mechanical, chemical and thermal energy. Differences in energy levels between one place and another will cause

Table 8.1 Groundwater Compositions (mg/l except pH) [209].

Solute	Basalt	Tuff	Shale	Granite
Alkalinity (as CaCO$_3$)	146	98	530	140
Calcium	<0.1	10	100	300
Iron	0.3	0.007	0.01	0.007
Magnesium	<1	3	50	3
Manganese	0.005	0.001	0.3	0.03
Potassium	3.0	4.2	24	2.3
Sodium	300	50	700	300
Strontium	0.01	0.05	3	5
Silica	100	70	10	10
Chloride	140	7	61	73
Fluoride	52	2.3	0.1	0.8
Phosphate	0.1	<0.01	0.02	0.02
Sulphate	75	19	2000	980
pH	9.3	7.8	8.4	8.3

Table 8.2 Groundwater composition (mg/l except pH) before and after 180-day glass leach [209].

Water	Condition (°C)	Fluoride	Sodium	Calcium	Magnesium	Iron	SiO$_2$	pH
Basalt	25 Before	53	260	1.5	0.07	0.190	110	9.3
	25 After	53	270	3.0	0.6	0.170	110	9.4
	90 Before	55	280	1.6	0.07	0.210	120	9.6
	90 After	55	290	7.2	2.0	4.40	150	9.6
Shale	25 Before	–	940	50	56	<0.009	14	8.4
	25 After	–	960	51	57	0.030	13	8.8
	90 Before	–	1000	10	40	<0.009	0.5	9.2
	90 After	–	1000	18	19	0.010	0.9	9.3
Tuff	25 Before	–	46	13	2.1	<0.009	65	7.8
	25 After	–	50	13	2.3	0.080	61	8.4
	90 Before	–	49	1.4	0.09	<0.009	60	8.6
	90 After	–	72	9.4	0.7	0.140	120	9.6

groundwater to migrate, where this is physically possible, to even out the distribution of energy. Pryce [210] has examined the movement of groundwater at depths of up to 5 km. It is his view that the general movement of deep groundwater is dominated by the general trend of the surface topography within a few kilometres of the area of interest, and that except in their immediate vicinity variations of structural properties of the rock have only a minor effect on the general character of flow. The laws which govern the movement of groundwater in porous media are now quite well understood but our knowledge of flow in fracture-dominated rock is still being developed. For a detailed review of how radionuclides move through rock in association with groundwater see the work of de Marsily *et al.* [211] and the results of recent *in situ* experiments in granites [212].

8.3.1 FLOW IN POROUS MEDIA

With porous media it is generally accepted that hydrogeological values for features such as porosity, gradient and hydraulic conductivity obtained at one point can often be extrapolated to cover a much larger volume of rock. What has come to be known as the elementary representative volume can be applied on the assumption that the rock mass is a continuous medium. Flow at this level has been defined by Darcy's law, which relates groundwater flow (also known as the filtration or Darcy velocity) to the gradient and the hydraulic conductivity of the particular rock medium through which water is moving. There are many ways of expressing Darcy's law, but at its simplest it states that the rate of flow per unit area of an aquifer is proportional to the gradient of the potential head measured in the direction of flow. This may be written as

$$v \propto i$$

where v is the specific velocity of the groundwater given in terms of length and time as cm s^{-1} or m d^{-1}, and i is the hydraulic gradient.

If we introduce a constant of proportionality, k, known as the hydraulic conductivity, which has the dimensions of length/time usually expressed as cm s^{-1}, m d^{-1} or as darcies (1 darcy = 0.831 m d^{-1} at 20°C, and 1 cm s^{-1} = 864 m d^{-1}), Darcy's law may then be written as

$$Q = kiA,$$

where Q = flow and A = cross-sectional area at right angles to the flow.

Typical values of hydraulic conductivity for fractured igneous and metamorphic rock would range between 10^{-2} and 10^{-6} cm s^{-1}.

Comparable values for unfractured material would be 10^{-8} to 10^{-11} cm s^{-1}, with shale exhibiting a very similar range.

The specific velocity, Q/A, is not the true velocity. The true velocity of groundwater movement in rock is invariably much greater than the specific velocity owing to the tortuous passage that individual droplets of water need to take around grains or along fractures which are far from straight. Darcy's law is only applicable to non-turbulent laminar-flow regimes. The Reynolds number, which takes account of the fluid density, velocity, viscosity, and the diameter of the pores through which flow takes place effectively, determines whether the flow will be laminar or not. Turbulence in groundwater flow is difficult to detect, but in general Reynolds numbers above 10 will result in non-laminar turbulent flow. In open fissures and under high hydraulic gradients Darcy's law may not therefore apply. It is also important to consider whether Darcy's law is applicable to very small hydraulic gradients. While in sandy media there may be no lower limit to the applicability of Darcy's law, for clay media there is apparently a limiting gradient below which the permeability is zero [211]. These are matters which, along with the calculation of convective forces, will require careful consideration, especially if a clay repository is to be proposed.

8.3.2   FLOW IN FISSURED AND FRACTURED MEDIA

The analysis of flow in fissured rock is approached in the literature from two directions based upon continuum and non-continuum models. If fracture density is extremely low or particular fissures are of interest it may be possible or indeed necessary to analyse flow in individual fissures as reviewed by Witthe [213], i.e. on a non-continuum basis. The continuum model based upon the work of Snow [214, 215] and Gale [73] involves treating the rock mass and its included fractures as if it were a large-grained porous medium and involves the replacement of the fractured rock by a representative continuum to which spatially defined values of hydraulic conductivity, porosity and compressibility can be assigned. This approach averages the hydrogeological properties over a large scale. Freeze and Cherry [96] have argued that this is a valid approach so long as the fracture spacing is sufficiently dense that the fractured medium acts in a hydraulically similar manner to a granular porous medium. The conceptualization is the same, although the representative elementary volume is considerably larger for fractured media than for granular media. Figure 8.2 portrays the equivalent concept approach in diagrammatic form, which in this example sees flow through apertures of 0.0034 cm spaced 1 m apart equating to flow through a porous medium with a hydraulic conductivity of 3.3×10^{-6} cm s^{-1}.

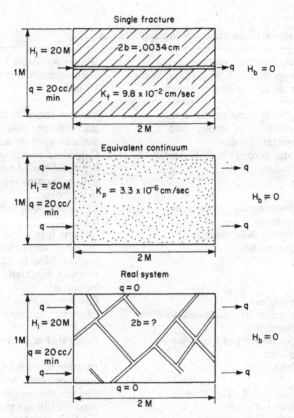

Figure 8.2 The equivalent continuum concept. $2b$=fracture width, K_f=
fracture permeability, K_p=intergranular permeability, q=flow,
H_1=head one, and H_b=head two [216].

To be applicable the equivalent concept requires laminar-flow
conditions to prevail in the rock–groundwater system of interest.

Bailey and Marine [217] have produced equations to estimate flow
in individual fractures. Flow (Q) is given as:

$$Q = \frac{2}{3}\left[\frac{(P_o - P_d)\,d_B^3\,d_w}{v d_L}\right].$$

where

$\quad Q$=flow, $d^3 t^{-1}$,
$\quad P_o$=pressure at inlet of fracture, $M d^{-1} t^{-2}$,
$\quad P_d$=pressure at outlet of fracture, $M d^{-1} t^{-2}$,
$\quad d_B$=half-width of fracture,
$\quad d_w$=length of fracture perpendicular to direction of flow,
$\quad v$=viscosity of liquid, $M d^{-1} t^{-1}$ (see Figure 8.1),
$\quad d_L$=length of fracture from inlet to outlet,
$(P_o - P_d)/d_L$=hydraulic gradient,

Table 8.3 Summary of log applications [219].

Required Information on the Properties of Rocks, Fluid, Wells, or the Groundwater System	Widely Available Logging Techniques Which Might be Utilized
Lithology and stratigraphic correlation of aquifers and associated rocks	Electric, sonic, or caliper logs made in open holes. Nuclear logs made in open or cased holes
Total porosity or bulk density	Calibrated sonic logs in open holes, calibrated neutron or gamma-gamma logs in open or cased holes
Effective porosity or true resistivity	Calibrated long-normal resistivity logs
Clay or shale content	Gamma logs
Permeability	No direct measurement by logging. May be related to porosity, injectivity, sonic amplitude
Secondary permeability – fractures, solution openings	Caliper, sonic, or borehole televiewer or television logs
Specific yield of unconfined aquifers	Calibrated neutron logs
Grain size	Possible relation to formation factor derived from electric logs
Location of water level or saturated zones	Electric, temperature or fluid conductivity in open hole or inside casing. Neutron or gamma-gamma logs in open hole or outside casing
Moisture content	Calibrated neutron logs
Infiltration	Time-interval neutron logs under special circumstances or radioactive tracers
Direction, velocity, and path of groundwater flow	Single-well tracer techniques – point dilution and single-well pulse. Multiwell tracer techniques
Dispersion, dilution, and movement of waste	Fluid conductivity and temperature logs, gamma logs for some radioactive wastes, fluid sampler
Source and movement of water in a well	Injectivity profile. Flowmeter or tracer logging during pumping or injection. Temperature logs
Chemical and physical characteristics of water, including salinity, temperature, density, and viscosity	Calibrated fluid conductivity and temperature in the well. Neutron chloride logging outside casing. Multielectrode resistivity

Table 8.3 Summary of log applications [219].

Required Information on the Properties of Rocks, Fluid, Wells, or the Groundwater System	Widely Available Logging Techniques Which Might be Utilized
Determining construction of existing wells, diameter and position of casing, perforations, screens	Gamma-gamma, caliper, collar, and perforation locator, borehole television
Guide to screen setting	All logs providing data on the lithology, water-bearing characteristics, and correlation and thickness of aquifers
Cementing	Caliper, temperature, gamma-gamma. Acoustic for cement bond
Casing corrosion	Under some conditions caliper, or collar locator
Casing leaks and/or plugged screen	Tracer and flowmeter

and the velocity of liquid through the fracture as function of distance from the fracture centreline is given as:

$$V_x \left[\frac{(P_o - P_d)}{2vd_L} \right] [1 - (d_x^2/d_B)],$$

where

d_x = distance from centreline of fracture

The average groundwater velocity in the fracture is two-thirds of the centreline velocity.

8.4 PROBLEMS IN DEFINING THE RELEVANT HYDROGEOLOGICAL PARAMETERS

It is not so much the production of an adequate mathematical treatment of groundwater flow, however, that poses problems for the hydrogeologist assessing HLW disposal sites, but rather the provision of enough accurate rock and groundwater data to feed into the available mathematical models and expressions. Problems can also arise in properly conceptualizing flow in rocks, especially in fractured media which will generally be the case for rock repositories. Abelin *et al.* [218] have carried out a series of detailed hydrogeological experiments at the Stripa Mine in Sweden and have discovered that fractures there are not uniform, being closed in some places and of irregular width in others, that flow may take place through only limited parts of a fracture giving rise to channelling, and that some groundwater, even though it is connected to fissures, may be effectively stagnant. While the injection of sorbing tracers can in part define where flow is taking place in such fracture-dominated systems, it cannot as yet provide a comprehensive flow analysis. Yet it remains important to be able to identify zones of channelling, as by concentrating flow this reduces the amount of contact surface between the flowing water and the bedrock and so reduces the opportunity for the host rock to absorb radionuclides. Zones of diffusion, in contrast, enhance the ability of the bedrock to sorb and retard radionuclides.

Standard methods exist to determine the hydrogeological parameters of fractured and porous media as shown in Table 8.3, and these are now being added to by new techniques such as radar [220–222]. The reliability and applicability of individual methods of determining hydrogeological properties can still pose problems, however. Table 8.4, for example, shows average values for hydraulic conductivity for four boreholes sunk 300 m into granite in Cornwall, in the United Kingdom. It can be seen that whilst there are a number of broadly comparable sets of results from the three different testing methods used, there are also some marked differences. It has been suggested in this case that the main differences between the 30-day pumping test figures and the other two test methods can be accounted for by such factors as the storage effects of the boreholes, which had a radius of 0.08 m. Nevertheless, results such as these highlight the need

Table 8.4 Comparison of results from various hydrogeological testing methods [223].

	Borehole A	Borehole B	Borehole C	Borehole D
Constant rate injection tests				
Average hydraulic conductivity $(m\ s^{-1} \times 10^{-10})$	11.2	15.0	3.01	3.47
Transient abstraction tests				
Average hydraulic conductivity $(m\ s^{-1} \times 10^{-10})$	11.7	12.3	8.91	4.17
30 day pumping test				
Average hydraulic conductivity $(m\ s^{-1} \times 10^{-10})$	48.5	24.7	pumping borehole	53.2

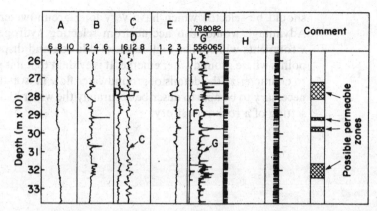

Figure 8.3 Suite of logs from an experimental test site for radioactive waste storage in granite. (a) Temperature (°C), (b) gamma (counts s^{-1}), (c) resistivity (unscaled), (d) neutron (counts s^{-1}), (e) density (counts s^{-1}), (f) caliper (mm), (g) sonic (m s^{-1}), (h) core log, (i) TV log. (Modified from [221].)

to assess the hydrogeological characteristics of any particular site by using as many different approaches as possible, while at the same time being aware of the strengths and shortcomings of each. This is clearly shown in Fig. 8.3 which gives a suite of logs using different techniques, but all intended to identify zones of permeability in a Swedish granite. Care must also be taken, of course, that the drilling of boreholes to gain the necessary characterization of the repository rock block does not prejudice its ultimate integrity by providing new flow paths back to the biosphere.

While it is as a liquid phase that groundwater generally occurs, it is also necessary to consider its effects on a repository as a solid in existing or anticipated permafrost areas, or as a gas should temperatures around a repository rise to sufficiently high levels given the prevailing pressure conditions.

8.5 CONCLUSIONS

Transportation by groundwater potentially represents the greatest single threat to the integrity of a HLW rock repository. While the ingress of groundwater can be reduced by the use of canisters and buffer and backfill materials, and the leachability of the waste can be reduced by solidification, it would be prudent to assume that groundwater will eventually come into contact with the waste. In the event that radionuclides are mobilized in groundwater it will be necessary to have an idea of groundwater flow paths and rates. Sites should be selected which have very long groundwater flow paths. Advantage would also accrue from selecting hydrogeological environments which offer high levels of dilution and dispersion at the points where groundwater emerge. It is evident that it is not sufficient to characterize the details of groundwater flow in isolation. It is also necessary to be able to describe accurately the wider hydrogeological setting of a rock repository.

9 Risk assessment and release scenarios for rock repositories

9.1 INTRODUCTION

It is clearly a matter of the utmost importance to have an idea of how any particular repository will operate in the years following final sealing of the HLW. It is to be expected that those responsible for final repository site selection and licensing will settle upon geological formations with characteristics which will offer a high degree of predictability and containment. To achieve this it is not sufficient to assess just the present situation but it is necessary to speculate as to how a repository's physical characteristics may change over time. These changes may broadly be categorized under three headings: those related to changes of a geological nature such as volcanism or glaciations, those concerning man-derived activity such as drilling into a repository by mistake, and those brought about by the interaction of the HLW and the repository barriers, both natural and engineered. Consequent upon any changes taking place in the initial repository conditions it is then necessary to assess how these changes may affect the ability of the repository to retain radionuclides or satisfactorily retard their migration.

9.2 APPROACHES TO RELEASE SCENARIOS AND RISK ASSESSMENT

A number of concepts have been developed to deal with the problems of future repository performance. Scenario analysis [224] involves the investigation of how the isolation system's containment barriers are breached by the combined action of natural, human-, and waste-initiated phenomena. Consequence analysis then seeks to describe the post-failure behaviour [225]. While it is useful to have these two conceptual ideas it is not necessarily wise to pursue them in isolation from one another. It is helpful, perhaps, to consider these concepts as essentially posing two questions: does the repository scenario adopted describe unacceptable consequences, and if so can they be eliminated or reduced to acceptable limits by changing the design of the repository in some way [226]?

According to Burkholder [225] the development of scenario and consequence analysis can be approached on either a generic (general) or a system-specific (site-specific) basis. However, as Barr, Bingham and Tierney [226] point out , it is not entirely clear exactly what is meant by a generic scenario. They ask if a generic scenario is one that is common to all repositories and burial media, but conclude that it is not. To illustrate this point they refer to some scenarios that are possible only at certain sites e.g. disruption by faults and some types of human intrusion, but which are nonetheless termed 'generic' because they were developed without reference to any particular site. Clearly, in the early days of repository design it was necessary to produce a list of all those phenomena which might lead to repository failure. A generic approach to scenario analysis was valid at that time as it provided an early set of criteria against which to assess potential repository environments. As Tables 9.1 and 9.2 show, what are believed to be comprehensive lists of possible breaching mechanisms have now been drawn up. It would be wrong to suggest that every possible cause of repository failure has been thought of; nevertheless, it is likely that only a few remain to be identified.

The generic approach still has a role to play in improving our understanding of processes which are fundamental to the use of a rock

Table 9.1   Possible modes of failure in the seabed disposal concept [158].

Mode of failure

(i) It could be physically disrupted by mass movement or slumping of sediments, erosion, dissolution, faulting, folding, volcanic or seismic activity.

(ii) Thermal convection of soft sediments could be initiated if high temperatures are engendered by the waste itself.

(iii) Radioactive nuclides leached from the wasteform could be carried to the seafloor by active migration of pore water. This would be enhanced if high thermal gradients occur around the wasteform, and would be facilitated by the presence of fissures and other water-conducting strata. The effect could be reduced by adsorption of nuclides onto the sediment particles.

(iv) Even in the absence of pore-water migration, radionuclides could reach the seabed by diffusion through stationary pore-water. This could be slowed by adsorption of nuclides onto sediment particles.

(v) The medium could be physically weakened by the emplacement procedure, or physically or chemically altered by the heat or radioactivity of the waste (for example, remineralization might occur, so that minerals with desirable properties such as high adsorptivity are replaced by minerals with less desirable properties).

(vi) Naturally occurring gas, or gas released from solid hydrates by radiogenic heat, could disrupt the medium.

(vii) Safe emplacement might be difficult or impossible because of the presence of obstructions such as boulders, or because of other unsuitable geotechnical properties.

Table 9.2 Phenomena potentially relevant to release scenarios for waste repositories [227].

Natural Processes and Events

Climatic change
Hydrology change
Sea level change
Denudation
Stream erosion
Glacial erosion
Flooding
Sedimentation
Diagenesis
Diapirism
Faulting/seismism
Geochemical changes

Fluid interactions
   Groundwater flow
   Dissolution
   Brine pockets

Uplift/subsidence
   Orogenic
   Epeirogenic
   Isostatic

Undetected features
   Faults, shear zones
   Breccia pipes
   Lava tubes
   Intrusive dykes
   Gas or brine pockets

Magmatic activity
   Intrusive
   Extrusive

Meteorite impact

Human Activities

Undetected past intrusion
   Undiscovered boreholes
   Mine shafts
Improper design
   Shaft seal failure
   Exploration borehole
   seal failure
Improper operation
   Improper waste
   emplacement
Transport agent
   introduction
   Irrigation
   Reservoirs
   Intentional artificial
   groundwater recharge
   or withdrawal
   Chemical liquid waste
   disposal

Climate control

Large scale alterations of hydrology
Intentional intrusion
   War
   Sabotage
   Waste recovery

Inadvertent future intrusion
   Exploratory drilling
   Archeological exhumation
   Resource mining (mineral, water,
     hydrocarbon, geothermal, salt, etc.)

Waste and Repository Effects

Thermal effects
   Differential elastic
   response
   Non-elastic response
   Fluid pressure, density,
   viscosity, changes
   Fluid migration
Chemical effects
   Corrosion
   Waste package
     – Geology interactions
   Gas generation
   Geochemical alterations

Mechanical effects
   Canister movement
   Local fracturing
Radiological effects
   Material property changes
   Radiolysis
   Decay product gas generation
   Nuclear criticality

repository, for instance in describing how groundwater moves through fractured rock at depth or how various earth materials actually adsorb radionuclides. Nevertheless, ultimately it is necessary to show how the general body of knowledge relates to any particular repository environment and this will necessitate many new system-specific studies.

Three main scenario-analysis methods have been widely used and described in the literature; they all take essentially the same information but treat it in different ways. In the expert-opinion method, which according to Burkholder [225] is the most widely used, those with a high degree of knowledge of the phenomena involved express a qualitative judgement on any particular repository situation. While possible, it is not usual for there to be a quantitative assessment of the experts' level of confidence in their judgement. This, together with the initial reliance upon opinion, albeit argued and reasoned, represents a weakness in the expert-opinion approach to scenario and consequence analysis.

A more quantitative approach is embodied in the fault-tree method of analysis [228]. This involves breaking down the repository into its various components, initially into a schematic diagram such as that given by D'Alessandro and Bonne [228] in Fig. 9.1. These individual but still large components can then be broken down in turn into smaller events and systems. Events are organized and analysed according to traditional fault–event-tree methods which provide a quantitative risk analysis. The method describes the behaviour at a repository as an event or series of events leading to breaching of the

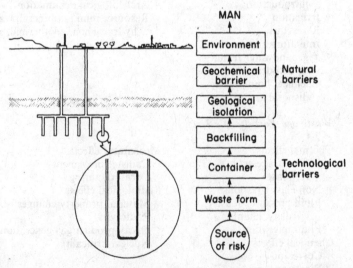

Figure 9.1　A simple model of the geological repository concept [228].

repository. Fault-tree analysis starts with careful definition of the failure events and systematically diagrams backwards to identify the events or combination of events that could cause the failure event to occur [229]. The potential repository failure is displayed graphically using logic in a tree-like structure, as shown in Fig. 9.2, which describes how radionuclides might be released to groundwater at the Mol Boom Clay site in Belgium. Event-tree analysis reverses the fault-tree approach by starting with the basic initiating events and working forwards in time to display their logical propagation to system failure events. By this method event trees diagrammatically illustrate the alternative outcomes or consequences of specified initiating events [229].

Clearly, in order to work the fault–event-tree method requires an absolute measure of the probability that any one component in the system will fail if an overall assessment of the chance of failure is to be computed. These data come from historical information about the past behaviour of similar systems. It is in the provision of these data that the fault-tree method is vulnerable to criticism. Many geologists hold the view that geological events cannot be predicted with any great degree of certainty [230, 231]. Look, for example, at the difficulties of anticipating the occurrence of earthquakes and volcanic eruptions. While it may be true that nature and geologic events are essentially random except at the detailed level where they are clearly governed by laws, it is possible to apply a degree of probability of any event's occurring at a particular place. For example, volcanic activity is unlikely to occur in the United Kingdom during the necessary lifetime of a repository, and glaciation is unlikely in low-lying tropical regions over a similar time period. Each event could happen, but the probability of its occurring is very low or almost zero in the context of our present state of knowledge. If faulting were taken as another case in point, this would be far more difficult to assess in any particular situation. Clearly, there will be tectonically active areas where faulting is very likely to occur, and at the other extreme there will be very stable areas where new faulting is very unlikely. Then there will be all the intervening situations of fault occurrence, and so it can be seen that a degree of probability of any event's occurring can be made, albeit the degree of confidence that can be placed upon such assessments will vary from scenario to scenario. It seems, to the author at least, that while fault-tree analysis is capable of providing a measure of quantitative assessment it too, like the expert opinion method, is to a large extent reliant upon qualitative judgements, especially where historical data are non-existent or patchy.

The other method of scenario analysis which has been used is known as the simulation method. This approach describes a specific repository system as a set of continuous processes upon which

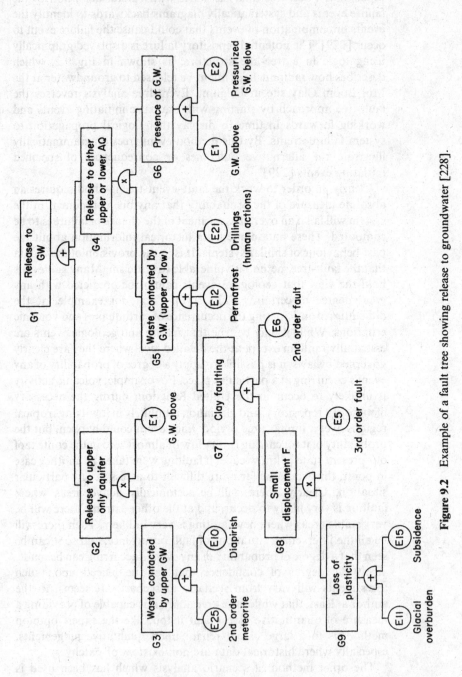

Figure 9.2 Example of a fault tree showing release to groundwater [228].

random discrete events are superimposed. This approach is particularly suited to the analysis of discrete stochastic events. The system relies for its effectiveness upon the description of accurate models to describe the various repository processes. Various data sets are fed into the models which are then run to see what effect changing parameters has on the overall integrity of the repository. Once specific repository sites have been chosen and detailed rock and design data are available this approach to repository safety should become more important and play a significant role in refining individual repository designs.

Various forms of statistical analysis are available to help the geologists in carrying out simulations, such as Monte Carlo/Markov chain analysis, the main applications and advantages of which are described in reference [229]. It is, however, beyond the scope of this work to describe these essentially mathematical techniques.

While these various predictive techniques all use different approaches to a common problem they all involve modelling of the waste repository to a certain extent. Models such as that shown in Fig. 9.3 are invaluable in concentrating effort at the main points of potential failure.

9.3 A TIME SCALE FOR PREDICTIONS

D'Alessandro and Bonne [228] have suggested that four time spans should be considered when examining the long-term performance of repositories. From the work of other authors, such as Zellmer [232] who has used computer simulation models with some 460 time steps covering a million-year period, the arguments behind D'Alessandro and Bonne's choice of time periods seems sound. The four time spans which they consider are:

a 2000 yr span, which they argue is representative of the historical time period and covers a period when geological phenomena have few possibilities to occur, and is sufficiently long to allow most fission products to decay completely;

a 25 000 yr span, which is extended enough to allow a few important climatic effects such as glaciation and permafrost to gain significance;

a 100 000 yr span, which may be long enough to cover an entire ice age, while other geological events of some importance may also start to affect the repository;

a 250 000 yr span, after which time the radiotoxicity of the waste is quite low and beyond which reliable geological predictions become increasingly difficult to make.

It may be appropriate to add a fifth much shorter time period of

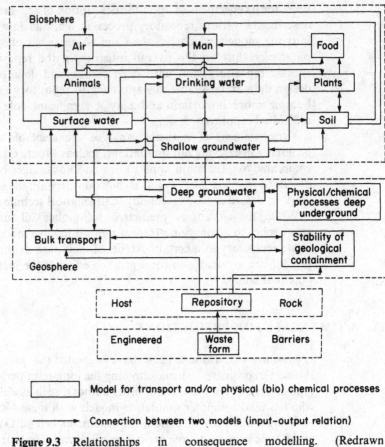

Model for transport and/or physical (bio) chemical processes

Connection between two models (input–output relation)

Figure 9.3 Relationships in consequence modelling. (Redrawn from [229].)

between 50 and 100 yr to cover the period during which a repository may be open and therefore subject to different risks.

9.4 EXAMPLES OF GENERIC AND SITE-SPECIFIC RELEASE SCENARIOS

To date, as Burkholder [225] has demonstrated, many release scenario projects using a multitude of different methods have been undertaken. It is clearly not possible to refer to them all. Nevertheless, an understanding of the contribution such studies make can be achieved by reference to illustrative examples.

9.4.1 THE GRISWOLD SCENARIO FOR THE WASTE ISOLATION
PILOT PLANT IN SOUTHEASTERN NEW MEXICO

Named after the geologist who first described it, this scenario has been further elaborated by Barr, Bingham and Tierney [226]. It is based upon an assumed salt dissolution threat to an evaporite repository. The repository tentatively proposed for this site is assumed to be located at a depth of about 700 m in the 500 m thick Salado Evaporite Formation which is itself part of an overall thickness of some 1200 m of evaporites [30]. The Salado Formation has a number of features which make it attractive as a potential HLW disposal site, including extremely low porosity and almost nil hydraulic conductivity. This, taken together with an almost total lack of brine pockets, suggests that the waste would be isolated hydrogeologically. Geochronological studies [233, 234] suggest that, as might be expected, these evaporites have not apparently been affected by dissolution or reprecipitation to any marked extent since their initial formation some 225 to 235 million years ago. Nevertheless, a release scenario based upon a large ingress of groundwater has been considered for this site.

The Griswold scenario is based on the presence of extensive potash deposits which have been worked some 200 m or so above the projected repository. Below these evaporite beds is the Capitan Reef which has been developed as an aquifer. Boreholes tapping the aquifer of necessity pass through the evaporite sequences which overlie it. Griswold has postulated that at some time in the future the casings which at present protect the water wells where they pass through the potash will corrode and water will then be able to enter the old mine workings. The scenario argues that groundwater moving towards low discharge points in old mine adits to the west will migrate through the old mines located above the proposed repository. The threat to the repository is that water will eventually remove the 100 to 200 m of evaporites between the old workings and the repository. The event tree depicting the scenario is given in Fig. 9.4.

It would be unrealistic to exclude this potential repository site solely on the basis of such speculation. If the scenario approach is to be of value it must be able to take a set of circumstances such as those described by Griswold and determine whether they will in practice pose a threat. This has of course been carefully carried out in this case. The point where fresh water might enter the old mine workings is some 8 km or so distant. Water travelling through evaporites over such a distance before passing over the repository would be more or less saturated with salts by the time it reached the evaporite beds above the repository. The potential of this salt-saturated groundwater to dissolve the salts above the repository would therefore be

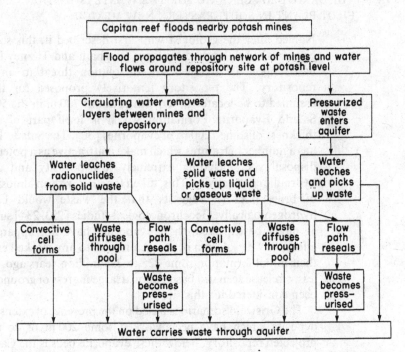

Figure 9.4 Event tree depicting the Griswold scenario for the Waste Isolation Plant. (Redrawn from [226].)

almost nil. Further, with a concentration of salts of about 400 000 mg l^{-1} the density of the brine would be such that flow movements would be very small, and it would be unable to reach the old mine adits which are too high in the circumstances.

Other groundwater intrusion scenarios have also been considered for the same repository site. These include models which link the overlying Rustler aquifer and an underlying aquifer in the Delaware Mountain Group [235]. This is shown in Fig. 9.5. Yet another release scenario for this repository envisaged a flow of groundwater from the Rustler Formation through the repository and then back to the Rustler Formation, as shown in Fig. 9.6. In both these cases initially anticipated flow paths based solely on head data were shown to be over-pessimistic when fluid density considerations were taken into account. When these are taken into account it seems that Delaware Mountain Group groundwater cannot flow up into the Rustler Formation and that fully saturated brine cannot flow down from the Rustler Formation into the repository workings.

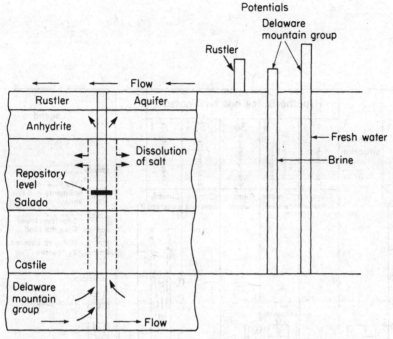

Figure 9.5 Schematic representation of the two-aquifer scenario for the Waste Isolation Pilot Plant. (Redrawn from [226].)

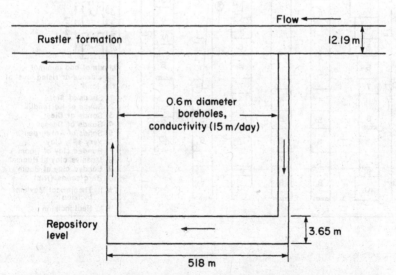

Figure 9.6 Schematic representation of the one-aquifer scenario for the Waste Isolation Pilot Plant. (Modified from [226].)

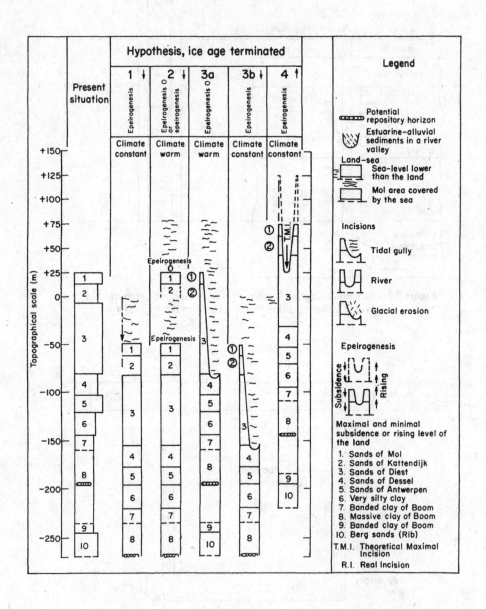

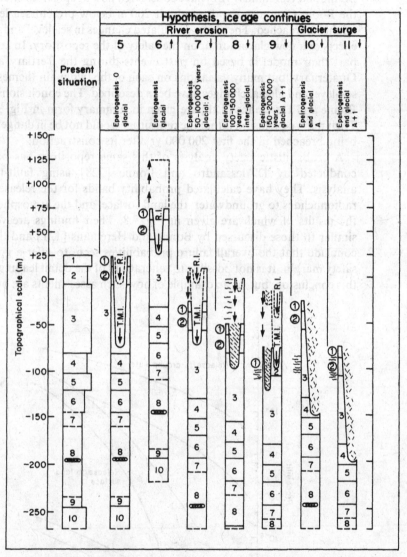

Figure 9.7 Some possible scenarios concerning future evolution of the potential repository at Mol over the next 2×10^5 years. (Redrawn and modified from [121].)

9.4.2 RELEASE SCENARIOS APPLIED TO THE BOOM CLAY FORMATION OF
BELGIUM

Bonne and Heremans [121] have considered how a repository sited in
the Boom Clay at a depth of about 200 m below present sea level
might be breached. They have postulated changes in sea level and the
effects of a new glaciation upon the safety of the repository. In large
part their model is based on past events during the Tertiary and
Quaternary, but many variations on each of the two main themes of
sea-level and glacial changes have been described. The conclusions of
Bonne and Heremans [121] are given in summary form in Fig. 9.7.
Their main conclusion is that the repository would not be in danger of
being breached in the first 200 000 yr after its construction.

A probabilistic safety analysis of this same repository has been
conducted by D'Alessandro and Bonne [228] using fault-tree
analysis. They have calculated probability bands for the release of
radionuclides to groundwater, the land surface, and the atmosphere,
the results of which are given in Fig. 9.8. Their findings are very
similar to those discussed by Bonne and Heremans [121] and they
conclude that the overall failure probabilities seem to offer a good
safety margin. It is not possible to reiterate all of the detail leading to
this conclusion, but, as an example of how such assessments are built

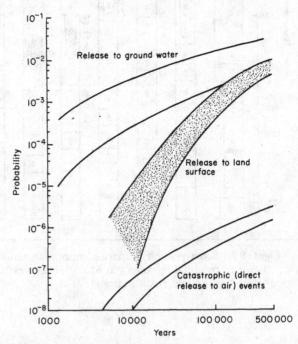

Figure 9.8 Top event probability bands [228].

up and arrived at, the way these two authors have assessed the impact of faulting is worthy of examination.

They began with the notion that the probability per year of faulting occurring (P_0) was equal to the present number of faults per square kilometre (N), divided by the time through which these faults occurred (T). N, calculated from *Landsat* imagery, gave 568.5 km of faults in the study area. T was not so readily computed, it being difficult to distinguish between new faults and remobilization of pre-existing ones. However, it is likely that the main period of faulting coincided with the major part of rift dislocations which occurred between 30 and 40 million years ago in association with the Rhone–Rhine Rift. This would give a T value of 30 to 40 million years. In the absence of decisive data to the effect that more recent faulting has, however, not taken place a much lower and inherently safer figure of 2 million years was eventually adopted as the appropriate T value. Thus the fault frequency for the area under consideration was calculated to be equal to 5×10^{-9} faults per year per square kilometre. By multiplying this figure by a further figure defined by the mean fault length in the area [236], and after taking into account the ability of clay to absorb faulting to a certain extent, a probability value for an offset fault originating and cutting through the repository in such a way as to be able to breach the clay formation of $1\text{–}5 \times 10^{-8}$ per year [228] is arrived at.

9.4.3 RELEASE SCENARIO STUDIES AT THE BASALT WASTE ISOLATION PROJECT, PASCO BASIN, WASHINGTON STATE, UNITED STATES OF AMERICA

The geology and repository concept envisaged in the Columbia River Basalts has already been described in §6.6.1. Arnett *et al.* [237] have examined potential failure lists of the type given in Tables 9.1 and 9.2. From these they have produced a list of release scenarios they believe to be relevant to the Basalt Waste Isolation Project. They readily rejected a number of release scenarios as being so remote as to be of little relevance. For example, the probability that a meteorite impact could excavate a repository some 1000 m below ground level over the one-million-year lifetime of the repository is estimated as 5×10^{-13} per year. The depth of the repository is such that most surface phenomena, with the exception of glaciations and climatic changes and their subsequent alteration of hydrologic factors, can also be discounted. The authors finally settled upon the following list of preliminary release scenarios:

interconnecting fault,
magmatic dike or sill intrusion,

Table 9.3 Scope of analysis with respect to a release scenario in a theoretical granite [238].

Scope of risk analysis

(1) Preliminary risk analysis of a probably typical concept of repository in granite formation of Japan.
(2) Analysis of risk in post-closure period, i.e. from about 1000 years after emplacement to a million years;
(3) Identification of dominant natural phenomena as well as design parameters which may contribute to the risk.

Stages in risk analysis

(1) Preparation of the waste disposal system concept;
(2) Identification of dominant steady state release paths of the radionuclides from repository to biosphere in post-closure period (reference release paths);
(3) Estimation of consequence due to reference release paths;
(4) Identification of events that may affect the reference release paths;
(5) Estimation of probability of such events;
(6) Estimation of consequence due to reference release paths modified by such events;
(7) Evaluation of risk;
(8) Safety research and development recommendation.
Several feedback processes from the steps (2) to (7) were planned to utilize available efforts and resources efficiently.

Table 9.4 Reference waste disposal system concept characteristics [238].

Characteristics assumed*

(1) The repository is located in granite formation at a depth of about 1 000 m below the surface;
(2) Major faults may exist at a distance of 2 km or more away from the centre of the repository;
(3) The geological media are saturated with the groundwater;
(4) The groundwater velocity is not very low even in the deep geologic formation;
(5) The hydraulic gradient depends principally on the geographical features around the site;
(6) The surface water flow (e.g. a river) exits near the site;
(7) The sea is not far from the site.
(8) The site is located in the area without any records of significant volcanic activities.

*These characteristics were assumed for a reference waste disposal system concept shown in Figure 9.8. It is based on the multiple barrier concept, considering the hydrogeological structure and geographical features of Japan.

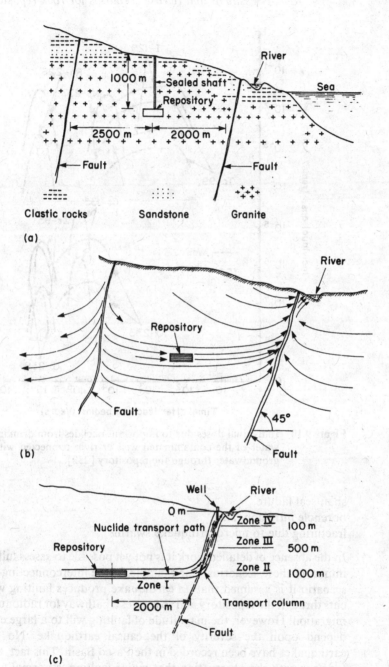

Figure 9.9 (a) Conceptual waste disposal system model in a Japanese granite. (b) Pattern of the regional groundwater flow through the conceptual granite repository. (c) Model showing the nuclide transport paths from the conceptual granite repository to the biosphere [238].

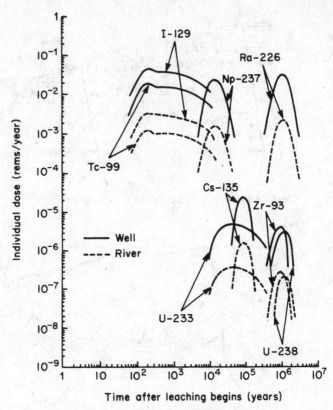

Figure 9.10   Individual doses due to important nuclides from drinking the water of the contaminated well or river connected with the groundwater through the repository [238].

shaft seal failure,
borehole intrusion, and
fracturing due to microearthquake swarms

In the absence of detailed work it is not yet possible to assess fully the impact of these scenarios. For example, in the interconnecting-fault scenario it is assumed that an earthquake produces faulting which cuts through the repository and so creates a pathway for radionuclide migration. However, the magnitude of faulting will to a large extent depend upon the severity of the causal earthquake. No large earthquakes have been recorded in the Pasco Basin. This fact, taken together with the observation that major faulting is normally only associated with earthquake magnitudes of 7 to 8 which are generally confined to plate boundaries (none of which pass near to the repository area), suggests that this may not be a realistic scenario either. Nonetheless, a methodical approach of the type already begun

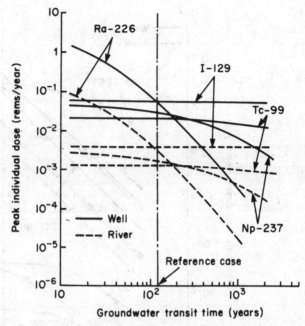

Figure 9.11 Relation between peak individual doses due to important nuclides and groundwater transit time, other parameters being unchanged [238].

is capable of concentrating research where it is most needed. Indeed it is the view of Arnett *et al.* [237] that the basic selection criteria for the identification of release scenarios should consist of an assessment of credibility, the probability of a significant release, and the degree of consistency with the latest site-specific data and knowledge.

9.4.4 A RISK ANALYSIS OF HLW DISPOSAL IN GRANITE

Kondo *et al.* [238] have conducted a risk analysis of HLW disposal into a typical Japanese granite. The procedure they adopted is broken down as shown in Table 9.3. They assumed a number of specific characteristics for their theoretical granite repository as shown in Table 9.4 and Fig. 9.9, in effect producing a hybrid site-specific–generic approach. With reference to a number of further assumptions such as that the waste container will be disrupted within 1000 yr, and an appraisal of retardation and release-enhancing effects with respect to radionuclides, a set of predicted results as shown in Figs 9.10 and 9.11 were produced. The value of such an approach is that it lends itself to speculative modelling; for example, if the leaching rate can be reduced what effect would this have on the peak individual doses per

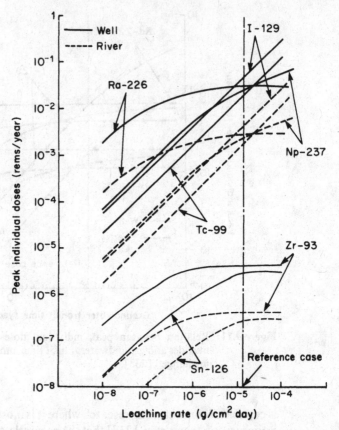

Figure 9.12 Relation between individual doses due to important nuclides and leaching rate, other parameters being unchanged [238].

year? The answer is given in Fig. 9.12. Such data can guide the search for suitable granites as well as refining repository design.

On a more dramatic plane than leaching by groundwater Kondo *et al.* [238] also estimated the effect of the repository being totally disrupted by a volcanic explosion, resulting in the sudden release of radionuclides to the atmosphere. The calculated amounts of fallout are given in Fig. 9.13.

9.5 CONCLUSIONS REGARDING RELEASE SCENARIOS AND RISK ANALYSIS

The early work carried out was of a generic nature, which for the most part confined itself to the examination of idealized models. While such an approach has its limitations it has served to identify the main

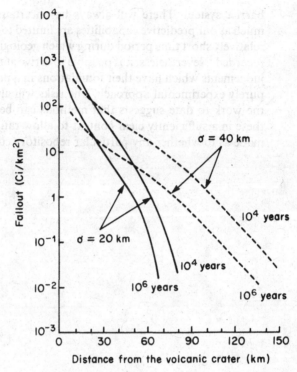

Figure 9.13 Fallout due to volcanic explosion through a conceptual Japanese granite repository at 10^4 years and 10^6 years after disposal respectively. The distribution of fallout is assumed to be Gaussian [238].

causes of repository failure and has highlighted those areas where further research is required to decide whether certain options are safe. A general picture has been built up of the possible radiological consequences to man of adopting a geological disposal option, and the initial results have been encouraging enough to prompt a second phase of study. This involves the detailed examination of specific sites to describe properly their relevant physical, chemical, hydrological and cultural features. Models can then be constructed to show how these various areas of interest interact. Various waste disposal strategies can then be superimposed upon these models and so the likely reaction of the repository can be gauged.

Masure and Venet [239] have provided a comprehensive review of progress to date and work still to be done when considering the disposal of HLW in continental geological formations. They have emphasized the unpredictability of certain aspects of the earth sciences and have warned against being misled by apparently very refined and precise work which relates to only part of the whole multi-barrier system. There will always be uncertainty in geology in as much as our predictive capabilities are limited to a large extent by the relatively short time period during which geological events have been recorded. Nevertheless, it is possible to arrive at reasoned and argued judgements which have their foundations in a pragmatic as well as a purely experimental approach. The risks will always be present but the work to date suggests that methods can be developed to place these in a sufficiently clear context to allow rational decisions to be made as to whether any particular repository concept is acceptable.

References

The following abbreviations are used throughout the reference list: AECL: Atomic Energy of Canada Ltd; AERE: Atomic Energy Research Establishment; CEC: Commission of the European Communities; EPA: Environmental Protection Agency; HMSO: Her Majesty's Stationery Office; IAEA: International Atomic Energy Agency; ICRP: International Commission on Radiological Protection; IMO: International Maritime Organization; IOS: Institute of Oceanographic Studies; NEA: Nuclear Energy Agency; NRC: Nuclear Regulatory Commission; NRPB: National Radiological Protection Board; OECD: Organization for Economic Cooperation and Development; ONWI: Office of Nuclear Waste Isolation; ORNL: Oak Ridge National Laboratories; UKAEA: United Kingdom Atomic Energy Authority; UN: United Nations; USDOE: United States Department of Energy; USGS: United States Geological Survey.

1. NEA (1977) *Objectives, Concepts and Strategies for the Management of Radioactive Waste Arising from Nuclear Power Programmes.* Report by a group of experts of the OECD, Paris.
2. Ferguson, J. E. (1982) *Inorganic Chemistry and the Earth*, Pergamon Press, Oxford.
3. Blomke, J. O. (1976) Management of radioactive wastes. In *Nuclear Power Safety* (eds J. H. Rust and L. E. Weaver), Pergamon Press, Oxford.
4. IAEA (1981) *Underground Disposal of Radioactive Wastes, Basic Guidance.* IAEA Safety Series No. 54, Vienna.
5. Flowers, R. H., Roberts, L. E. J. and Tymons, B. J. (1986) Characteristics and quantities of radioactive wastes. *Philos. Trans. R. Soc. London, Ser. A*, **319**, 5–16.
6. Pentreath, R. J. (1980) *Nuclear Power, Man and the Environment*, Taylor and Francis, London.
7. NEA (1984) *Geological Disposal of Radioactive Waste*, OECD, Paris.
8. NEA (1981) *Safety of the Nuclear Fuel Cycle*, OECD, Paris.
9. Ewart, F. T. (1982) *Fuel Inventories and Derived Parameters for the Reactor Systems CDFR, LWR, AGR and Magnox.* Report AERE-R10037, UKAEA, AERE, Harwell, UK.
10. Brookins, D. G. (1984) *Geochemical Aspects of Radioactive Waste Disposal*, Springer-Verlag, New York.
11. Bocola, W. (1983) *Characterization of Long-Lived Radioactive Waste to be Disposed in Geological Formations, Parts 1 and 2.* Working document SCK/CEN, Mol, Belgium.
12. Wang, T. S. Y., Tsang, C. F., Cook, N. G. W. and Witherspoon, P. A. (1981) Long-term thermohydrologic behaviour of nuclear waste

repositories. In *Predictive Geology* (eds G. de Marsily and D. F. Merriam), Pergamon Press, Oxford.

13. NEA (1982) *Nuclear Energy and its Fuel Cycle, Prospects to 2025*, OECD, Paris.

14. Barnes, R. W. (1979) The management of irradiated fuel in Canada. In Paper 79-10 of the Canadian Geological Survey (ed. C. R. Barnes).

15. NEA (1982) *Geological Disposal of Radioactive Waste: Geochemical Processes*, OECD, Paris.

16. UN (1977) *Report to the United Nations Scientific Committee on the Effects of Atomic Radiation*. General Assembly Document 32, Session Supplement No. 40(A/31/40), United Nations, New York.

17. Clark, M. J. and Kelly, G. N. (1981) The radiation exposure of the UK population from nuclear industry effluents. In: *The Environmental Impact of Nuclear Power*, British Nuclear Energy Society, London.

18. ICRP (1959) *Recommendations of the International Commission on Radiological Protection*. ICRP Publication No. 2.

19. Papp, T. (1986) The role of the canister in a system for the final disposal of spent fuel or high-level waste. *Philos. Trans. R. Soc. London, Ser. A*, **319**, 39–47.

20. NEA (1984) *Long-term Radiation Protection Objectives in Radioactive Waste Disposal*. Report by a group of experts of the OECD, Paris.

21. Department of the Environment (1984) *Disposal Facilities on Land for Low- and Intermediate-Level Radioactive Wastes: Principles for the Protection of the Human Environment*, HMSO, London.

22. Gera, F. (1981) Geologic predictions and radioactive waste disposal: a time limit for the predictive requirements. In *Predictive Geology*, (eds G. de Marsily and D. F. Merriam), Pergamon Press, Oxford.

23. Hamstra, J. and Van der Feer, Y. (1981) New ICRP norms and underground disposal of radioactive wastes. *Energie Spectrum* 81, 4, ECN, The Netherlands (in Dutch).

24. Zurkinden, A. and Niederer, U. (1980) Release scenarios: water intrusion as a normal case. In *Radionuclide Release Scenarios for Geologic Repositories*. Proceedings of the NEA Workshop, Paris, 8–12 September 1980.

25. Matthess, G. (1982) *The Properties of Groundwater*, John Wiley and Sons, New York and Chichester.

26. Chapman, N. and McEwen, T. (1986) Geological solutions for nuclear wastes. *New Scientist*, 28 August 1986, pp. 36–40.

27. Zeller, E. J., Saunders, D. F. and Angino, E. E. (1973) *Science and Public Affairs*, **29**, 1, 4.

28. Mather, J. D., Greenwood, D. A. and Greenwood, P. B. (1979) Burying Britain's radioactive waste – the geological areas under investigation. *Nature*, **2814**, 332–334.

29. IAEA (1977) *Site Selection Factors for Repositories of Solid High-Level and Alpha-Bearing Wastes in Geological Formations*. Tech. Rep. No. 177, IAEA, Vienna.

30. *Geological Characterisation Report, Waste Isolation Pilot Plant [WIPP] Site, Southeastern New Mexico* (1978). In *Report* AT(29-1)-789 (2 vols) (eds D. W. Powers, S. J. Lambert, S. E. Shaffer, L. R. Hill and W. D. Weart), USDOE, Washington, DC.

31. Chapman, N. A. (1979) *Geochemical Considerations in the Choice of a Host Rock for the Disposal of High-Level Radioactive Wastes*, Institute of Geological Sciences. HMSO, London.

32. Usiglio, J. (1849) Analyse de l'eau de la Mediterranée sur les côtes de France. *Ann. Chem.*, **27**, 92–107, 172–191.

33. Stewart, F. H. (1963) *Marine evaporites*. United States Geological Society, Professional Paper No. 440-Y.

34. Borchet, H. and Muir, R. O. (1964) *Salt Deposits: the Origin, Metamorphism and Deformation of Evaporites*, D. Van Nostrand Co. Ltd, London.

35. Jackson, M. P. A. and Talbot, C. J. (1986) External shapes, strain rates, and dynamics of salt structures. *Geol. Soc. Am. Bull.* **97**, 305–323.

36. Office of Nuclear Waste Isolation (1981) *Geologic Evaluation of Gulf Coast Salt Domes: Overall Assessment of the Gulf Interior Region*. Report No. 106.

37. Physikalisch–Technische Bundesanstalt (1983) *Zusammenfassender Zwischenbericht uber Bisherige Ergebnisse der Standortuntersuchung in Gorleben*, Braunschweig.

38. Pedersen, A. and Lindstrom, J. (1980) Selection of release scenarios for a Danish waste repository in a salt dome. In *Radionuclide Release Scenarios for Geologic Repositories*. Proceedings of the NEA Workshop, Paris, 8–12 September 1980.

39. Clairborne, H. C. (1972) *Neutron-Induced Transmutation of High-Level Radioactive Waste*. ORNL-TM-3964, Oak Ridge, Tennessee.

40. Bell, F. G. (1983) *Engineering Properties of Soils and Rocks*, Butterworths, London.

41. Ode, H. (1969) *Review of Mechanical Properties of Salt Relating to Salt-Dome Genesis in Diapirism and Diapirs. Am. Assoc. Petroleum Geol.*, *Memoir* 8.

42. Baar, C. A. (1977) The *in situ* behaviour of salt rocks. In *Applied Salt Rock Mechanics*, Elsevier, New York.

43. Bradshaw, R. L. and McClain, W. C. (1971) *Project Salt Vault: A Demonstration of the Disposal of High-Activity Solidified Wastes in Underground Salt Mines*. Paper No. ORNL-4555, Oak Ridge, Tennessee.

44. Jockwer, N. (1981) Transport phenomena of water and gas components within rock salt in the temperature field of disposed high-level waste. In *Near-Field Phenomena in Geologic Repositories for Radioactive Waste*, NEA, OECD Paris.

45. Bachwan, G. O. and Johnson, R. (1974) *Stability of Salt in the Permian Salt Basin of Kansas, Oklahoma, Texas and New Mexico*. Open File Report 74-194, US Geological Survey, Washington DC.

46. Piper, A. M. (1973) *Subrosion in and about the Four-Township Study Area near Carlsbad, New Mexico*. ORNL Subcontract 3745, Oak Ridge, Tennessee.

47. James, A. N. and Lupton, A. R. R. (1978) Gypsum and anhydrite in foundations of hydraulic structures. *Geotechnique*, **28**, 249–273.

48. Martinez, J. D. (1981) Salt domes – the past and present, keys to the future. In *Predictive Geology* (eds G. de Marsily and D. F. Merriam), Pergamon Press, Oxford.

49. Kent, P. E. (1979) The emergent Hormuz salt plugs of southern Iran. *J. Petroleum Geol.*, **2,** 117–144.
50. Kupfer, D. H. (1976) Time and rates of salt movement in north Louisiana. In *Salt-Dome Utilization and Environmental Considerations, a Symposium* (eds J. D. Martinez and R. L. Thoms), pp. 145–170, Institute of Environmental Studies, Louisiana State University.
51. Vita-Finzi, C. (1979) Rates of Holocene folding in the coastal Zagros near Bandar Abbas, Iran. *Nature*, **278,** 632–633.
52. Pfiffner, O. A. and Ramsay, J. G. (1982) Constraints on geological strain rates: arguments from finite strain states of naturally deformed rocks. *J. Geophys. Res.*, **87,** 311–321.
53. Jenks, G. H. (1975) *Gamma-Irradiation Effects in Geologic Formations of Interest in Waste Disposal*. ORNL-TM-4827, Oak Ridge, Tennessee.
54. Jenks, G. H. and Bopp, C. D. (1974) *Storage and Release of Radiation Energy in Salt in Radioactive Waste Repositories*. ORNL-TM-4449, Oak Ridge, Tennessee.
55. Hatch, F. H., Wells, A. K. and Wells, M. K. (1961) *Petrology of the Igneous Rocks*, George Allen and Unwin, London.
56. Siever, R. (1982) *Earth*, 3rd edn, W. H. Freeman and Co., San Francisco.
57. Allard, B., Larson, S. A., Albinsson, Y., Tullborg, E. L. *et al.* (1981) Minerals and precipitates in fractures and their effects on the retention of radionuclides in crystalline rocks. In *Near-Field Phenomena in Geologic Repositories for Radioactive Waste*, NEA, OECD, Paris.
58. Blacic, J. D. (1981) Importance of creep failure of hard rock in the near field of a nuclear waste repository. In *Near-Field Phenomena in Geologic Repositories for Radioactive Waste*, NEA, OECD, Paris.
59. Hustrulid, W. (1980) Crystalline rock mining technology. In *Technical Progress Report for the Quarter Oct.* 1 to *Dec.* 31, 1979, *ONWI*, 9(5), Office of Nuclear Waste Isolation, Battelle Memorial Institute, Columbus, Ohio.
60. Barbreau, A., Bonnet, M., Goblet, P., Margat, J. *et al.* (1981) Long-term prediction of the fate of nuclear waste deeply buried in granite. In *Predictive Geology* (eds G. de Marsily and D. F. Merriam), Pergamon Press, Oxford.
61. Brotzen, O. (1981) Predictive geology in nuclear-waste management. In *Predictive Geology* (eds G. de Marsily and D. F. Merriam), Pergamon Press, Oxford.
62. Davis, S. N. and Turk, L. J. (1964) Optimum depths of wells in crystalline rocks. *Ground Water*, **2,** 6–11.
63. Neretnieks, I. (1981) The influence of microfissures in crystalline rock on radionuclide migration. In *Predictive Geology* (eds G. de Marsily and D. F. Merriam), Pergamon Press, Oxford.
64. Grundfelt, B. (1978) *Nuclide Migration from a Rock Repository for Spent Fuel*. Kemakta Konsult AB KBS (Stockholm), Technical Report No. 77.
65. NEA (1983) *Geophysical Investigations in Connection with Geological Disposal of Radioactive Waste*. Proceedings of a workshop held in Ottawa, September 1982, AECL, OECD, Paris.

66. Fritz, F., Barker, J. F. and Gale, J. (1983) Isotope hydrology at the Stripa test site. In *In situ Experiments in Granite for Geological Disposal of Radioactive Waste*. Proceedings of a workshop held in Stockholm, October 1982, OECD, Paris.

67. Davison, C. C. and Simmons, G. R. (1983) The research program at the Canadian Underground Research Laboratory. In *In situ Experiments in Granite for Geological Disposal of Radioactive Waste*. Proceedings of a workshop held in Stockholm, October 1982, OECD, Paris.

68. Carlsson, L., Norlander, H. and Olsson, T. (1983) Hydrogeological investigations in boreholes, In *In situ Experiments in Granite for Geological Disposal of Radioactive Waste*. Proceedings of a workshop held in Stockholm, October 1982, OECD, Paris.

69. Hultberg, B., Larson, S. A. and Tullborg, E. L. (1981) *Grundvatten i Kristallin Berggrund*. SGU Dnr 41.41.-81-H206-U, Uppsala.

70. Giletti, B., Siever, R., Handin, J., Lyons, J. and Pinder, G. (1978) *State of Geological Knowledge Regarding Potential Transport of High-Level Radioactive Waste from Deep Continental Repositories*. EPA/520/4-78-004, US Environmental Protection Agency, Washington, DC.

71. Neretnieks, I. (1979) Analysis of some tracer runs in granite rock using a fissure model. In *Scientific Basis for Nuclear Waste Management*, Vol. I, Plenum Press, New York.

72. Gale, J. E., Witherspoon, P. A., Wilson, C. R. and Rouleau, A. (1983) Hydrogeological characterization of the Stripa site. In *In situ Experiments in Granite for Geological Disposal of Radioactive Waste*. Proceedings of a workshop held in Stockholm, October 1982, OECD, Paris.

73. Gale, J. E. (1982) *Assessing the Permeability Characteristics of Fractured Rock*. Geol. Soc. Am. Special Paper No. 189.

74. Rankama, K. (1963) *Progress in Isotope Geology*, John Wiley–Interscience, London.

75. Jenks, G. H. (1975) *Gamma-Irradiation Effects in Geologic Formations of Interest in Waste Disposal: A Review and Analysis of Available Information and Suggestions for Additional Experimentation*. ORNL-TM-4827, Oak Ridge, Tennessee.

76. Benson, L. V., Carnahan, C. L., Apps. J. A., Mouton, C. A., Corrigan, C. J., Frisch, C. J. and Shomura, L. K. (1980) *Basalt Alteration and Basalt–Waste Interaction in the Pasco Basin of Washington State*. US DOE Report W8A-SBB-51760, Washington, DC.

77. Pettijohn, E. J. (1975) *Sedimentary Rocks*, Harper and Row, London.

78. Sykes, M. L. and Smyth, J. R. (1980) Location-specific studies, Yucca Mountain. In *Evaluation of Tuff as a Medium for Nuclear Waste Repository; Interim Status Report on the Properties of Tuff* (eds J. K. Johnstone and K. Wolfsberg), Scandia National Laboratory Report SAND80-1464, Albuquerque, New Mexico.

79. Parrish, D. K., Waldman, H. and Osnes, J. D. (1981) Temperatures and stresses in the vicinity of a nuclear waste repository in welded tuff. In *Near-Field Phenomena in Geologic Repositories for Radioactive Waste*, NEA, OECD, Paris.

80. Johnstone, J. K., Sundberg, W. D. and Krumhansl, J. L. (1981) A

parametric study of the effects of thermal environment on a waste package for a tuff repository. In *Near-Field Phenomena in Geologic Repositories for Radioactive Waste*, NEA, OECD, Paris.

81. Potter, P. E., Maynard, J. B. and Pryor, W. A. (1984) *Sedimentology of Shale. Study Guide and Reference Source*, Springer-Verlag, New York.

82. Deer, W. A., Howie, R. A. and Zussman, J. (1969) *An Introduction to the Rock-Forming Minerals*, Longmans Green, London.

83. Stanley, D. J. (ed.) (1972) *The Mediterranean Sea, a Natural Sedimentation Laboratory*, Dowden Hutchinson and Ross, Stroudsberg, PA.

84. Whitten, D. G. A. and Brooks, J. R. V. (1974) *The Penguin Dictionary of Geology*, Penguin Books, Harmondsworth, England.

85. Keller, W. D. (1964) Processes of origin and alteration of clay minerals. In *Soil Clay Mineralogy: a Symposium* (eds C. I. Rich and G. W. Kunze), University of North Carolina Press, Chapel Hill, North Carolina.

86. Heremans, R., Barbreau, A., Bourke, P. and Gies, H. (1980) Thermal aspects associated with the disposal of waste in deep geological formations. In *Radioactive Waste Management*. Proceedings of a CEC seminar held in Luxembourg, 1980, Harwood Academic Publishers, Brussels.

87. Tassoni, E. (1980) An experiment on the heat transmission in a clay rock. In *Proceedings of NEA Workshop on Use of Argillaceous Materials for the Isolation of Radioactive Waste*. Proceedings of a workshop held in Paris, September, 1979, OECD, Paris.

88. Heremans, R., Buyens, M. and Manfroy, P. (1978) Le comportement de l'argle vis-à-vis de la chaleur. In *In situ Heating Experiments in Geological Formations*. Proceedings of a seminar held in Ludvika/Stripa, September 1978, OECD, Paris.

89. Heremans, R., Bonne, A. and Manfroy, P. (1981) Experimentation on, and evaluation of near-field phenomena in clay: The Belgian approach. In *Near-field Phenomena in Geologic Repositories for Radioactive Waste*. NEA, OECD, Paris.

90. Leoni, L., Polizzano, C., Sartori, F. and Sensi, I. (1984) Chemical and mineralogical transformations induced in Pliocenic clays by a small subvolcanic body and consequences for the storage of radioactive wastes. *Neues Jahrbuch für Mineralogie–Monatshefte, Jahrgang* 1984.

91. Polizzano, A. (1983) *Etudes sur les effets de la Chaleur Naturelle dans les Argiles*. Note presented during an AIPEA Congress at Prague, September 1983.

92. Aoyagi, K., Kobayashi, N. and Kazama, T. (1975) *Clay Mineral Facies in Argillaceous Rocks of Japan and their Sedimentary Petrological Meanings*. Proceedings of an international clay conference, Applied Publishing, Illinois.

93. Hedberg, H. D. (1936) Gravitational compaction of clays and shales. *Am. J. Sci. Ser.* 5, **31**, 1035–1072.

94. Conybeare, C. E. B. (1967) Influence of compaction on stratigraphic analysis. *Bull. Canadian Petroleum Geol.* **15**, 331–345.

95. Davis, S. N. (1969) Porosity and permeability of natural materials. In *Flow Through Porous Media* (ed. R. J. M. DeWiest), Academic Press, New York and London.

96. Freeze, A. R. and Cherry, J. A. (1978) *Groundwater*, Prentice Hall, Englewood Cliffs, New Jersey.

97. Skytte-Jensen, B. (1982) *Migration Phenomena of Radionuclides into the Geosphere*. EUR 7676. Commission of the European Communities, Harwood Academic Publishers, Brussels.

98. Ames, L. L. and Rai, D. (1978) *Radionuclide Interactions with Soil and Rock Media*. Vol. 1: *Processes Influencing Radionuclide Mobility and Retention; Element Chemistry and Geochemistry; Conclusions and Evaluation*. EPA Report 520/6-78-007A.

99. Muller, A. B., Langmuir, D. and Duda, L. E. (1983) The formation of an integrated physicochemical/hydrologic model for predicting waste nuclide retardation in geologic media. In *Scientific Basis for Nuclear Waste Management*, Vol. 6. Proceedings of a symposium held in Boston, MA, November 1982, North-Holland, New York.

100. Brookins, D. G. (1978) *Application of E_h–pH Diagrams to Problems of Retention and/or Migration of Fissiogenic Elements at Oklo*, IAEA Tech. Publ. No. 119.

101. Brookins, D. G. (1978) E_h–pH diagrams for elements from $Z = 40$ to $Z = 52$: application to the Oklo natural reactor. *Chemical Geology*, **23**, 325–342.

102. Brookins, D. G. (1978) Retention of transuranic and actinide elements and bismuth at the Oklo natural reactor, Gabon: Application of E_h–pH diagrams. *Chemical Geology*, **23**, 309–323.

103. Brookins, D. G. (1980) Syngenetic model for some early proterozoic uranium deposits: evidence from Oklo. In *International Uranium Symposium on the Pine Creek Geosyncline*, CSIRO Institute of Earth Resources and IAEA.

104. Laul, J. C. and Papike, J. J. (1982) *Chemical Migration by Contact Metamorphism Between Granite and Silt Carbonate Systems*. IAEA Symposium No. 257.

105. Williams, A. E. (1980) *Investigation of Oxygen-18 Depletion of Igneous Rocks and Ancient Meteoric–Hydrothermal Circulation in the Alamosa River Stock Region, Colorado*. Unpublished PhD thesis, Brown University.

106. Brookins, D. G. (1981) Geochemical study of a lamprophyre dike near the WIPP Site. In *The Scientific Basis for Nuclear Waste Management*, Vol. III (ed. J. G. Moore), Plenum Press, New York.

107. Warner, B. F. (1978) The storage in water of irradiated oxide fuel elements. Testimony presented at the Windscale Inquiry on Spent Fuel Reprocessing, Windscale, United Kingdom. HMSO, London.

108. Saunders, P. A. H. (1981) The management of high-level waste and its environmental impact. In *The Environmental Impact of Nuclear Power*, British Nuclear Energy Society, London.

109. Sombret, C. G. (1985) The vitrification of high-level radioactive wastes in France. *Nuclear Energy*, **24**, 85–98.

110. Mendel, J. E., Nelson, R. D., Turcotte, R. P., Gray, W. J. *et al.* (1981) *A State-of-the-Art Review of Materials Properties of Nuclear Waste Forms*. Battelle Pacific Northwest Laboratories Report PNL-3802-UC-70.

111. Ringwood, A. E., Kesson, S. E., Ware, N. G., Hibberson, W. and

Major, A. (1979) Immobilization of high-level nuclear wastes in SYNROC. *Nature*, **278**, 219.

112. Ringwood, A. E. and Kelly, P. M. (1986) Immobilization of high-level waste in ceramic waste forms. *Philos. Trans. R. Soc. London, Ser. A*, **319**, 63–82.

113. *Final Storage of Spent Nuclear Fuel* (1983), SKBF/KBS-3, Stockholm.

114. Moak, D. P. (1981) *Waste Package Materials Screening and Selection*. Batelle Mem. Inst. Report ONWI-312, Columbus, Ohio.

115. Dayal, R. *et al.* (1982) *Nuclear Waste Management Technical Support in Development of Nuclear Waste Form Criteria for the NRC*. Vol. I, NUREG/CR-2333, BNL NUREG 51458, Brookhaven National Laboratory, Brookhaven, New York.

116. Fyfe, W. S. and Haq, Z. (1979) Nuclear waste disposal: geochemical and other aspects. In Paper 79-10 of the Canadian Geological Survey (ed. C. R. Barnes).

117. NRC (1982) *Rationale for the Performance Objectives*. 1OCFR 60, Nuclear Regulatory Commission, Washington, DC.

118. Lester, D. H. (1981) Waste package performance assessment. In *Near-Field Phenomena in Geologic Repositories for Radioactive Waste*. NEA, OECD, Paris.

119. Nowak, E. J. (1980) The backfill as an engineered barrier for nuclear waste management. In *The Scientific Basis for Nuclear Waste Management*, Vol. II, (ed. C. J. M. Northrup), Plenum Press, New York.

120. Pitman, S. G. (1980) *Investigation of Susceptibility of Titanium Grade 2 and Titanium Grade 12 to Environmental Cracking in a Simulated Basalt Repository Environment*. PNL-3915, Battelle Pacific Northwest Laboratory, Richland, Washington.

121. Bonne, A. and Heremans, R. (1981) Scenarios d'evolution geologique lente, appliqués au site argileux de Mol, Belgique. In *Radionuclide Release Scenarios for Geologic Repositories*, NEA, OECD, Paris.

122. Klingsberg, C. and Duguid, J. (1980) *Status of Technology for Isolating High-Level Radioactive Wastes in Geologic Repositories*. EPA Report TIC11207.

123. Pusch, R. (1979) Highly compacted sodium bentonite for isolating rock deposited radioactive waste products. *Nuclear Technology*, **45**, 153–157.

124. Pusch, R. (1977) *Required Physical and Mechanical Properties of a Buffer Mass*. SKBF/KBS Tech. Rep. No. 33, Stockholm.

125. Radhakrishna, H. S. and Tsui, K. K. (1981) Thermal properties of buffer/backfill materials and their effects on the near-field thermal regime in a nuclear fuel waste disposal vault. In *Near-Field Phenomena in Geologic Repositories for Radioactive Waste*, NEA, OECD, Paris.

126. Komarnei, S. and Roy, R. (1980) Supercoverpack: tailor-made mixtures of zeolites and clays. In *The Scientific Basis for Nuclear Waste Management*, Vol. II, (ed. C. J. M. Northrup), Plenum Press, New York.

127. Nowak, E. J. (1979) *The Backfill as an Engineered Barrier for Nuclear Waste Management*. Sandia National Laboratories Rep. SAND79-0990C, Albuquerque, New Mexico.

128. Nowak, E. J. (1980) *Radionuclide Sorption and Migration Studies of*

Getters for Backfill Barriers. Sandia National Laboratories Rep. SAND-79-1110, Albuquerque, New Mexico.

129. Nowak, E. J. (1980) *The Backfill Barrier as a Component in a Multiple Barrier Nuclear Waste Isolation System.* Sandia National Laboratories Rep. SAND79-1305, Albuquerque, New Mexico.

130. Simpson, D. R. (1980) *Desiccant Materials Screening for Backfill in a Salt Repository.* ONWI-214, Office of Nuclear Waste Isolation, Battelle Memorial Institute, Columbus, Ohio.

131. Coons, W. E., Moore, E. L., Smith, M. J. and Kaser, J. D. (1980) *The Functions of an Engineered Barrier System for a Nuclear Waste Repository in Basalt.* DOE Report RHO-BWI-LD-23, Washington, DC.

132. CEC (1983) *Admissible Thermal Loading in Geological Formations: Consequences for Waste Management Methods.* Vol. 1, *Synthesis Report*, Vol. 2, *Crystalline Rocks*, Vol. 3, *Salt Formations*, Vol. 4, *Clay Formations*, CEC 8179, Luxembourg.

133. Scott, J. A. (1983) *Limits on the Thermal Energy Release from Radioactive Wastes in a Mined Geologic Repository*, ONI-4, Office of NWTS Integration, Battelle Memorial Institute, Columbus, Ohio.

134. Chan, T., Lang, P. A. and Thompson, P. M. (1985) Mechanical response of jointed granite during shaft sinking at the Canadian Underground Research Laboratory. In *Proceedings of the 26th US Symposium on Rock Mechanics – Rock Mechanics: Research and Engineering Applications in Rock Masses, Rapid City, South Dakota, USA, 26–28 June, 1985* (ed. E. Ashworth), Rotterdam: A. A. Balkema.

135. Sir William Halcrow and Partners (1981) *Repository Schemes for High-Level Radioactive Waste Disposal.* A review of schemes for argillaceous and saliferous formations, prepared for the Department of the Environment, UK.

136. Hustrulid, W. and Ubbes, W. (1983) Results and conclusions from rock mechanics/hydrology investigations: CSM/ONWI Test Site. In *In situ Experiments in Granite for Geological Disposal of Radioactive Waste.* Proceedings of a workshop held in Stockholm, October 1982, OECD, Paris.

137. Muirhead, I. R. and Glossop, L. G. (1968) Hard rock tunnelling machines. *Trans. Inst. Mining and Metallurgy, Sect. A, Mining Industry*, **77**, A1–21.

138. Pirie, M. D. (1972) The use of rock tunnel machines. *Proc. Inst. Civil Eng.* **57**, Part 1, paper 5755.

139. Bell, F. G. (1980) *Engineering Geology and Geotechnics*, Newnes-Butterworths, London.

140. Maisham, D. (1975) Ground freezing. In *Methods of Treatment of Unstable Ground* (ed. F. G. Bell), Newnes-Butterworths, London.

141. Baca, R. G., Langford, D. W. and England, R. L. (1981) Analysis of host rock performance for a nuclear waste repository using coupled flow and transport models. In *Near-Field Phenomena in Geologic Repositories for Radioactive Waste*, NEA, OECD, Paris.

142. Smith, M. J., Turner, D. A. and Deju, R. A. (1983) Repository and waste package designs for high-level nuclear waste disposal in basalt. In

Proceedings of the IAEA Conference on Radioactive Waste Management held in Seattle, WA, May 1983.

143. INFCE Study, Working Group 7/46 (1979) *A Repository for Solidified Nuclear Waste in a Deep Tertiary Clay Formation.* Prepared in support of the IAEA.

144. *Eignungsprüfung der Schachtanlage Konrad fur die Endlagerrung Radioaktiver Abfalle* (1982) GSF-T 136.

145. Burgess, A. S. and Sandstrom, P. O. (1980) *Irradiated Fuel and Immobilised Waste Vaults – Preliminary Design Concepts.* Technical Record TR-48, Atomic Energy of Canada Limited.

146. Davison, C. C. and Guvanasen, V. (1985) Hydrogeological characterisation, modelling and monitoring of the site of Canada's Underground Research Laboratory. In *Proceedings of the 17th Congress of the International Association of Hydrogeologists on Hydrogeology of Rocks of Low Permeability, Tuscon, Arizona, January 1985.*

147. Mather, J. D. and Sargent, F. P. (1986) Characteristics of crystalline rocks. *Philos. Trans. R. Soc. London, Ser. A,* **319**, 139–156.

148. Johnstone, J. K. and Wolfsberg, K. (eds) (1980) *Evaluation of Tuff as a Medium for Nuclear Waste Repository: Interim Status Report on the Properties of Tuff.* Sandia National Laboratory Rep. SAND80-1464, Albuquerque, New Mexico.

149. Spengler, R. W., Muller, D. C. and Livermore, R. B. (1979) *Preliminary Report on the Geology and Geophysics of Drill Hole UE 25a-1, Yucca Mountain, Nevada Test Site,* USGS Open File Report 79-1244.

150. Eberl, D. and Hower, J. (1976) Kinetics of illite formation. *Geol. Soc. Am. Bull.,* **87**, 1324–1330.

151. Moore, J. G., Godbee, H. W., Kibbey, A. H. and Joy, D. S. (1975) *Development of Cementitious Grouts for the Incorporation of Radioactive Wastes.* USDOE Rep. ORNL-4962, Oak Ridge, Tennessee.

152. Butlin, R. N. and Hills, D. L. (1982) *Backfilling and Sealing a Repository for High-Level Radioactive Waste: A Review,* CP 6/82, UK Building Research Station, Department of the Environment, London.

153. Mott, Hay and Anderson, Consulting Engineers (1982) *The Backfilling and Sealing of Radioactive Waste Repositories, Phase 1: Interim Report,* prepared for the CEC.

154. Christensen, G. L. (1980) Sadia borehole plugging for the Waste Isolation Pilot Plant. In *Borehole and Shaft Plugging,* Proceedings of a NEA workshop held in Columbus, Ohio, OECD, Paris.

155. Roy, D. M. (1980) Geochemical factors in borehole and shaft plugging materials stability. In *Borehole and Shaft Plugging,* Proceedings of a NEA workshop held in Columbus, Ohio, OECD, Paris.

156. Pusch, R. (1980) *Swelling Pressure of Highly Compacted Bentonite,* SKBF/KBS Tech. Rep. No. 80-13, Stockholm.

157. Pusch, R. and Borgesson, L. (1982) Preliminary results from the buffer mass test of phase I of the Stripa Project. In *In situ Experiments in Granite for Geological Disposal of Radioactive Waste.* Proceedings of a workshop held in Stockholm, OECD, Paris.

158. Searle, R. C. (1979) *Guidelines for the Selection of Sites for Disposal of Radioactive Waste on or Beneath the Ocean Floor.* IOS Report No. 91.

159. Holliday, F. G. T. (1984) *Report of the Independent Review of Disposal of Radioactive Waste in the Northeast Atlantic*, HMSO, London.

160. IMO (1972) *Convention on the Prevention of Marine Pollution by Dumping of Wastes and Other Matter, 29 December 1972*, London.

161. NEA (1984) *Seabed Disposal of High-Level Radioactive Waste*, OECD, Paris.

162. UN (1982) *United Nations Convention on the Law of the Sea*, A/CONF.62/122, 7 October 1982.

163. Ove Arup and Partners (1982) *Ocean Disposal of High-Level Radioactive Waste, Penetrator Option, Studies Relevant to Emplacement in Deep Ocean Sediments*, Department of the Environment Report No. DOE/RW/82.102, London.

164. Ove Arup and Partners (1983) *Ocean Disposal of High-Level Radioactive Waste, Penetrator Engineering Study, Phase 1, Preliminary Feasibility*, Department of the Environment Report No. DOE/RW/83.094, London.

165. Berger, W. H. (1974) Deep sea sedimentation. In *The Geology of the Continental Margins* (eds C. A. Burk and C. L. Drake), Springer Verlag, New York.

166. Reading, H. G. (ed.) (1978) *Sedimentary Environments and Facies*, Blackwell Scientific Publications, Oxford.

167. Keany, J., Ledbetter, M., Watkins, N., and Huang, T. C. (1976) Diachronous deposition of ice rafted debris in sub-Antarctic deep sea sediments. *Bull. Geol. Soc. Am.*, **87**, 873–882.

168. Rona, P. A. (1973) Worldwide unconformities in marine sediments related to eustatic changes in sea level. *Nature Phys. Sci.*, **244**, 26.

169. Kennett, J. P. and Watkins, N. D. (1976) Regional deep-sea dynamic processes recorded by Late Cenozoic sediments of the south-east Indian Ocean. *Bull. Geol. Soc. Am.*, **87**, 321–339.

170. Heezan, B. C. and Ewing, M. (1952) Turbidity currents and submarine slumps, and the 1929 Grand Banks earthquake. *Am. J. Sci.*, **250**, 849–873.

171. Eittreim, S. and Ewing, M. (1972) Suspended particulate matter in the deep waters of the North American Basin. In *Studies in Physical Oceanography* (ed. A. L. Gordon), Gordon and Breach, New York,

172. Kelling, G. and Stanley, D. J. (1976) Sedimentation in canyon, slope, and base of slope environments. In *Marine Sediment Transport and Environment Management* (eds. D. J. Stanley and D. J. P. Swift), John Wiley, New York.

173. IOS (1983) *CTD Data from the NE Atlantic 31N–46N, July 1982, Discovery Cruise 130* IOS Rep. No. 165.

174. Atkins, (1981) *Planning Concepts for the Disposal of High-Level Radioactive Waste: Task 7: the Deep Ocean Bed*, Department of the Environment Report No. DOE/RW/82.015, London.

175. Manschot, D., Callanfels, S. and Van, J. E. (1981) *A Survey of Emplacement Methods for Radioactive Canisters in Deep Ocean Sediments*. Marine consultants report to DORA, Holland.

176. Burchett, S. N. (1982) *Preliminary Penetrator Studies, Subseabed Disposal Programme, Annual Report*. SAND82-0664/1, Vol. II, Part 2, January–September 1981, Albuquerque, New Mexico.

177. Miller, M. M., Chryssostomidis, C. and Shirley, C. G. (1982) *Conceptual Ship Design, Subseabed Disposal Programme, Annual Report*. SAND82-0664/1, Vol. II, Part 2, January–September 1981, Albuquerque, New Mexico.

178. Young, C. W. (1981) An empirical equation for predicting penetration depth into soft sediments. *Ocean 81*, Vol. 2, pp. 674–677.

179. McNeil, R. L. (1981) Approximate method for estimating the strengths of cohesive materials from penetrator decelerations. *Ocean 81*, Vol. 2, pp. 688–693.

180. Freeman, T. J., Murray, C. N., Francis, T. J. G., McPhail, S. D. and Schultheiss, P. J. (1984) Modelling radioactive waste disposal by penetrator experiments in the abyssal Atlantic Ocean. *Nature*, **310**, 130–133.

181. Beard, R. M. (1981) A penetrator for deep ocean seafloor exploration. *Ocean 81*, Vol. 2, pp. 668–673.

182. Maddocks, D. V. and Savvidou, C. (1982) *Ocean Disposal of High-Level Radioactive Waste: Centrifuge Modelling of High-Level Radioactive Waste Placed in the Deep-Ocean Bed as a Projectile*. Progress report, 1 April–30 September 1982, Cambridge University Engineering Department, Cambridge, UK.

183. St John, H. D., Mayne, J. R. and Hills, D. L. (1983) *Ocean Disposal of High-Level Radioactive Waste, Backfilling and Sealing of Boreholes in the Deep Ocean Bed, Incorporating Containers of High-Level Radioactive Waste by Drilled Emplacement*. Department of the Environment Report No. DOE/RW/83.065, London.

184. Marples, J. A. C. *et al.* (1980) In *Radioactive Waste Management and Disposal* (eds R. Simon and S. Orlowski), Luxembourg.

185. Hughes, A. E. *et al.* (1981) Report R-10190, AERE Harwell, UK.

186. Braithwaite, J. W. and Magnani, N. J. (1979) *Subseabed Disposal Program Annual Report 1979*, SAND90-2577/II, Albuquerque, New Mexico.

187. Schaefer, D., Glass, R. S. and Abrego, L. (1980) *Subseabed Disposal Program Annual Report 1980*, SAND-1095/II, Albuquerque, New Mexico.

188. NEA (1981) *Proceedings of the Sixth Annual NEA Seabed Working Group Meeting*, SAND81-0427, Paris.

189. Marsh, G. P. *et al.* (1983) *Corrosion Assessment of Metal Overpacks for Radioactive Waste Disposal*, Department of the Environment Report No. DOE/RW/83.021, AERE, Harwell, UK.

190. Gartling, D. K. (1978) *COYOTE – A Finite-Element Computer Program for Nonlinear Heat Conduction Problems*. SAND77-1332, Sandia Laboratories, Albuquerque, New Mexico.

191. Gartling, D. K. (1979) *MARIAH – A Finite-Element Computer Program for Incompressible Porous Flow Problems. Part I: Theoretical Background*, SAND79-1662, Part II: *User's Manual*, SAND79-1623, Sandia Laboratories, Albuquerque, New Mexico.

192. Chavez, P. F. and Dawson, P. R. (1981) *Thermally Induced Motion of Marine Sediments Resulting from Disposal of Radioactive Wastes*. Sandia National Laboratories Rep. SAND-1476, Albuquerque, New Mexico.

193. Schreiner, F., Fried, S. and Friedma, A. (1981) A study of the mobility of plutonium and americium in marine sediments. In *Subseabed Disposal Program Annual Report, January to December 1980*, SAND81-1095/II(1), Albuquerque, New Mexico.

194. Schreiner, F., Fried, S. and Friedman, A. (1981) Diffusion of neptunyl(v) and pertechnetate ions in marine sediments. In *Subseabed Disposal Program Annual Report, January to September 1981*, SAND82-0664/II(1), Albuquerque, New Mexico.

195. Baetsle, L. H. and Mittempergher, M. (1980) Disposal in argillaceous formations. In *Radioactive Waste Management and Disposal* (eds R. Simon and S. Orlowski), Harwood Academic Publishers, Brussels.

196. Mott, Hay and Anderson (1984) *The Backfilling and Sealing of Radioactive Waste Repositories*, Vols. 1 and 2. EUR 9115, CEC, Luxembourg.

197. NEA (1981) *Research and Environmental Surveillance Programme Related to Sea Disposal of Radioactive Waste*, OECD, Paris.

198. Martin Marietta (1980) In *Proceedings of a Workshop on Physical Oceanography Related to the Subseabed Disposal of High-Level Nuclear Waste, Big Sky, Montana* (ed. A. R. Robinson) SAND80-1776, Albuquerque, New Mexico.

199. Pentreath, R. J. (1983) Biology. In *Interim Oceanographic Description of the North-East Atlantic Site for the Disposal of Radioactive Wastes*, OECD, Paris.

200. Mullin, M. M. and Robinson, A. R. (1981) Appendix A. In *Biological and Related Chemical Research Concerning Subseabed Disposal of High-Level Nuclear Waste: Report of a Workshop at Jackson Hole, Wyoming*, SAND81-0012, Albuquerque, New Mexico.

201. Needler, G. T. (1986) Dispersion in the ocean by physical, geochemical and biological processes. *Philos. Trans. R. Soc. London, Ser. A*, **319**, 177–189.

202. NRPB (1976) Publication No. NRPB-R48.

203. McGowan, J. A. (1974) The nature of oceanic ecosystems. In *The Biology of the Oceanic Pacific* (ed. C. B. Miller), Oregon State University Press, Corvallis.

204. IOS (1978) *Oceanography Related to Deep-Sea Waste Disposal*. IOS Report No. 77.

205. IOS (1985) *Status Report on Research at the Institute of Oceanographic Sciences Related to the Disposal of Radioactive Waste on or Beneath the Seafloor*. IOS Report No. 204.

206. Davis, S. N. (1982) *Hydrogeology of Radioactive Waste Isolation: the Challenge of a Rational Assessment*. Geol. Soc. Am. Special Paper No. 189.

207. Allard, B., Beall, G. W. and Krajewski, T. (1978) The sorption of actinides in igneous rocks. *Nuclear Technology*.

208. Brookins, D. G. (1979) Thermodynamic considerations underlying the migration of radionuclides in geomedia: Oklo and other examples. In *The Scientific Basis for Nuclear Management*, Vol. I (ed. G. J. McCarthy), Plenum Press, New York.

209. Rees, T. F., Cleveland, J. M. and Nash, K. L. (1985) Leaching of

plutonium from a radioactive waste glass by eight groundwaters from the western United States. *Nuclear Technology*, **70**, 133–140.

210. Pryce, M. H. L. (1986) Principles governing deep groundwater flow. *Philos. Trans. R. Soc. London, Ser. A*, **319**.

211. de Marsily, G., Goblet, P., Ledoux, E. and Barbreau, A. (1978) *Model of the Transfer of Dissolved Radioelements in Deep Geological Formations.* Fourth Seminar on Dispersion in Natural Media, French Society for Radiological Protection, Cadarache.

212. NEA (1985) *In situ Experiments in Granite.* Proceedings of the second NEA/Stripa Project symposium, OECD, Paris.

213. Witthe, W. (1973) General report on the symposium, 'Percolation Through Fissured Rock'. *Bull. Int. Assoc. Eng. Geol.*

214. Snow, D. T. (1968) Hydraulic characteristics of fractured metamorphic rocks of Front range and implications to the Rocky Mountain Arsenal Well. *Colorado School of Mines Quarterly*, **63**, 167–199.

215. Snow, D. T. (1969) Anisotropic permeability of fractured media. *Water Resources Research*, **5**, 1273–1289.

216. Webb, G. A. M. (1986) Radiological criteria for the disposal of radioactive wastes. *Philos. Trans. R. Soc. London, Ser. A*, **319**, 17–25.

217. Bailey, C. E. and Marine, I. W. (1980) *Parametric Study of Geohydrologic Performance Characteristics for Geologic Waste Repositories.* E. I. du Pont de Nemours and Company Report No. DP-15555.

218. Abelin, H., Moreno, L., Tunbrant, S., Gidlund, J. and Neretnieks, I. (1985) Flow and tracer movement in some natural fractures in the Stripa Granite. In *In situ Experiments in Granite*, OECD, Paris.

219. Fetter, C. W. (1980) *Applied Hydrogeology*, Charles E. Merrill, London.

220. Gringarten, A. C. (1982) *Flow-test Evaluation of Fractured Reservoirs.* Geol. Soc. Am. Special Paper No. 189.

221. Nelson, P. H. (1982) *Advances in Borehole Geophysics for Hydrology.* Geol. Soc. Am. Special paper No. 189.

222. Olsson, O., Forslund, O., Lundmark, L., Sandberg, E. and Falk, L. (1985) The design of a borehole radar system for detection of fracture zones. In *In situ Experiments in Granite*, OECD, Paris.

223. Feates, F. S., Margat, J. and Gray, D. A. (1980) Disposal of high-level radioactive waste into crystalline rocks. In *Radioactive Waste Management and Disposal* (eds R. Simon and S. Orlowski), Harwood Academic Publishers, Brussels.

224. IAEA (1980) *Safety Assessment for the Underground Disposal of Radioactive Wastes.* IAEA Draft Safety Series Report, Vienna.

225. Burkholder, H. C. (1981) The development of release scenarios for geologic nuclear waste repositories, where have we been? Where should we be going? In *Radionuclide Release Scenarios for Geologic Repositories*, OECD, Paris.

226. Barr, G. E., Bingham, F. W. and Tierney, M. S. (1981) The use and misuse of scenarios in waste-disposal studies. In *Radionuclide Release Scenarios for Geologic Repositories*, OECD, Paris.

227. Irish, E. R. (1981) Safety assessment for the underground disposal of radioactive waste – the IAEA programme. In *Radionuclide Release Scenarios for Geologic Repositories*, OECD, Paris.

228. D'Alessandro, M. and Bonne, A. (1981) Fault-tree analysis for probabilistic assessment of radioactive-waste segregation: an application to a plastic clay formation at a specific site. In *Predictive Geology* (eds G. de Marsily and D. F. Merriam), Pergamon Press, Oxford.

229. IAEA (1981) *Safety Assessment for the Underground Disposal of Radioactive Wastes*. IAEA Safety Series No. 56, Vienna.

230. Mann, C. J. (1970) Randomness in nature. *Geol. Soc. Am. Bull.* **81**, 95–104.

231. Kitts, D. B. (1976) Certainty and uncertainty in geology. *Am. J. Sci.* **276**, 29–46.

232. Zellmer, J. T. (1981) Computerized simulation of nuclear waste repositories in geologic media for release scenario development. In *Radionuclide Release Scenarios for Geologic Repositories*, OECD, Paris.

233. Schilling, G. A. (1973) K–Ar dates on Permian potash minerals from southeastern New Mexico. *Isochron/West*, No. 6.

234. Brookins, D. G. (1981) U–Pb dates for U(VI) hydrosilicates, Grants, New Mexico. *Isochron/West*, Vol. 32.

235. USDOE (1979) *Draft environmental impact statement – Waste Isolation Pilot Plant*, Vol. 1, Section 9.5, DOE/EIS-0026-D, Washington.

236. D'Alessandro, M., Murray, C. N., Bertozzi, G. and Girardi, F. (1980) Probability analysis of geological processes: a useful tool for the safety assessment of radioactive waste disposal. *Radioactive Waste Management*, Vol. 1, No. 1.

237. Arnett, R. C., Baca, R. G., Caggiano, J. A. and Carrell, D. J. (1981) Radionuclide release scenario selection process for a possible basalt repository. In *Radionuclide Release Scenarios for Geologic Repositories*, OECD, Paris.

238. Kondo, S., Tokushita, T., Hwang, M. J. and Murano, T. (1981) Risk analysis of high-level radioactive waste disposal into granite formation. In *Radionuclide Release Scenarios for Geologic Repositories*, OECD, Paris.

239. Masure, P. and Venet, P. (1980) Some considerations on radioactive waste disposal in continental geological formations. In *Radioactive waste management and disposal* (eds R. Simon and S. Orlowski), Harwood Academic Publishers, Brussels.

Adsorption The attraction and adhesion of a layer of ions from an aqueous solution to the solid mineral surfaces with which it is in contact.

Absorbed dose A measure of the energy deposited in matter by ionizing radiation. The special SI unit of absorbed dose is the gray. 1 gray (Gy) = 100 rad = 1 J kg^{-1}.

Actinide series The group of elements, from actinium (atomic number 89) to lawrencium (atomic number 103) which together occupy one position in the Periodic Table. They are all radioactive.

ALARA As low as reasonably achievable, economic and social factors being taken into account.

Activation products Nuclides which have become radioactive in consequence of bombardment by neutrons or other nuclear particles.

Alpha (α) particle A positively charged particle emitted in the decay of some radioactive nuclei, especially nuclides of the actinide series. It has very low penetrating power and hence pure α emitters are almost exclusively a hazard when taken into the body.

Aquifer Rock which is saturated with water and which is sufficiently permeable to allow its economic recovery via wells or springs.

BEIR Biological effects of ionizing radiation.

Beta (β) particle An electron or positron emitted in the decay of some radioactive nuclei. It is only moderately penetrating.

Breeder reactor A reactor that creates more fissionable fuel than it consumes. The new fissionable material is created by capture in fertile materials of neutrons from fission.

BWR Boiling water reactor.

Cladding Fuel elements in most nuclear reactors are made of fissile material clad in a protective metal sheathing which is highly resistant to the physicochemical conditions that prevail in a reactor core. The function of the cladding is to prevent corrosion of the fuel and escape of fission products to the coolant.

Collective dose equivalent (S) A quantity introduced to provide a measure of the health detriment in an exposed population. The unit of collective dose equivalent and its derivatives is the man sievert (man Sv).

Conditioning of waste Those operations that transform waste into a more manageable and stable form, such as vitrification or mixing with bitumen.

Consequence analysis A safety analysis that estimates potential individual and population radiation doses to humans based on radionuclide releases and transport from a waste repository to the human environment as defined by hypothetical release and transport scenarios.

Critical group A group as homogeneous as practicable with respect to radiation exposure, which is representative of the more highly exposed individuals in the population.

Daughter product A nuclide formed by the radioactive decay of another nuclide; a synonym for decay product.

Dose equivalent (*H*) A quantity which correlates better with the deleterious effect of exposure to ionizing radiation, more particularly with the delayed stochastic effects, than does absorbed dose. The special SI unit of dose equivalent and its derivatives is the sievert $(Sv) = 100 \text{ rem} = 1 \text{ J kg}^{-1}$.

Effective dose equivalent (*H*$_E$) An indicator of the stochastic risk assumed to result from any irradiation, whether uniform or non-uniform.

Event-tree analysis An inductive probabilistic technique that starts with hypothesizing the occurrence of basic initiating events and proceeds through their logical propagation to system failure events. The event tree is the diagrammatic illustration of the alternative consequences or outcomes of specified initiating events.

Fertile material A material, not itself fissionable by slow neutrons, that can be converted into a fissile material by irradiation in a reactor by capturing a neutron which is added to its nucleus. There are two basic important fertile materials, uranium-238 and thorium-232, which are partially converted to fissile plutonium-239 and uranium-233 respectively after neutron capture.

Fault-tree analysis A deductive probabilistic technique that starts with hypothesizing and defining failure events and systematically deduces the events or combinations of events that could cause the failure events to occur. The fault tree is the diagrammatic illustration of the events in a tree-like structure.

Fission-products Nuclides produced either by fission or by the subsequent radioactive decay of the radioactive nuclides thus formed.

Gamma (γ) rays Electromagnetic radiation with short wavelength (10^{-8} to 10^{-11} cm) which is emitted by the nucleus. The emission of γ rays accompanies the disintegration of many α or β emitters. The penetrating power of γ rays is a function of their energy.

GCR Gas-cooled reactor.

Generic analysis A generalized analysis for a repository as opposed to a site-specific study.

Groundwater The water contained in interconnected pores located below the water table or in a confined aquifer.

High-level waste The highly radioactive liquid, containing mainly fission products, as well as some actinides, which is separated during chemical reprocessing of irradiated fuel. The term also includes any other waste with radioactivity levels intense enough to generate significant quantities of heat by the radioactive decay process, together with spent reactor fuel.

HWR Heavy-water-moderated reactor.

Hydraulic conductivity A coefficient of proportionality describing the rate at which water can move through a porous medium.

IAEA International Atomic Energy Agency.

ICRP International Commission on Radiological Protection.

Intermediate-level waste Those wastes which, because of their radionuclide content, require shielding but need little or no provision for heat dissipation during their handling or disposal.

Ion exchange A process by which an ion in a mineral lattice is replaced by another ion which was present in an aqueous solution.

Isotope One of several nuclides with the same atomic number, but with different atomic mass, and hence differing in the number of neutrons.

Leaching This term is used in HLW disposal studies to describe the gradual erosion/dissolution of emplaced solid waste or chemicals therefrom, or the removal of sorbed material from the surface of a solid or porous bed.

Light-water reactor (LWR) A nuclear reactor that uses ordinary water as both a moderator and a coolant and utilizes slightly enriched uranium-235 fuel. There are two commercial LWR types: the boiling-water reactor (BWR) and the pressurized water reactor (PWR).

Low-level waste Those wastes which, because of their low radionuclide content, do not require shielding during normal handling and transportation.

Models In applied mathematics, the analytical or mathematical representation or quantification of a real system and the ways phenomena occur within that system. Individual or subsystem models can be combined to give system models.

Milling As used in uranium processing (concentration), a process in the uranium fuel cycle in which ore that contains only about 0.2% uranium oxide (U_3O_8) is concentrated into a compound called yellow cake, which contains 80–90% uranium oxide.

NEA Nuclear Energy Agency.

Overpack Secondary or additional external containment for packaged radioactive waste.

Porosity The ratio of the volume of void spaces in a rock to the total volume of the rock.

PWR Pressurized water reactor.

Radioactivity A process whereby certain nuclides undergo spontaneous disintegration in which energy is liberated, generally resulting in the formation of new nuclides. The process is accompanied by the emission of one or more types of radiation, such as α particles, β particles and photons (electromagnetic radiation).

Radiotoxic potential   This is derived from a summation of the comparative level of radioactivity of each radionuclide relative to the limit on its annual level of intake by ingestion recommended by the International Commission on Radiological Protection and a partial measure of the hazard posed to human health by radionuclides.

RCG   Radioactive concentration guide.

Reprocessing   A generic term for the chemical and mechanical processes used to remove fission products and recover fissile uranium-233, uranium-235 and plutonium-239, and fertile thorium-232, uranium-238 and other valuable material.

Scenario analyses   Safety analyses that identify and define phenomena, and their interactions, which could initiate and/or influence the release and transport of radionuclides from the source in a repository to humans.

Somatic effects   Health effects which become manifest in the exposed individual.

Stochastic effects   Those for which the probability of an effect occurring, rather than its severity, is regarded as a function of dose, without threshold.

Throw-away option   An alternative to reprocessing which involves discarding unreprocessed spent fuel.

Transmutation   The conversion of long-lived radionuclides into much shorter-lived or even stable non-radioactive nuclides.

Transuranium elements   Elements with atomic numbers greater than 92, including neptunium, plutonium, americium and curium.

Worst-case scenario   The scenario for release and transport of radionuclides from a waste repository to the biosphere which represents the most severe situation conceivable on the basis of pessimistic assumptions.

Index